Reassembling the UAW

Abe Walker

Reassembling the UAW

Insurgency, Contention, and the Struggle for Unionism in the American South

Temple University Press

Philadelphia • *Rome* • *Tokyo*

TEMPLE UNIVERSITY PRESS
Philadelphia, Pennsylvania 19122
tupress.temple.edu

Copyright © 2026 by Temple University Press—Of The Commonwealth System
of Higher Education
All rights reserved
Published 2026

Library of Congress Cataloging-in-Publication Data

Names: Walker, Abe, 1981– author
Title: Reassembling the UAW : insurgency, contention, and the struggle for
unionism in the American south / Abe Walker.
Description: Philadelphia : Temple University Press, 2026. | Includes
bibliographical references and index. | Summary: "This book describes
the organizational changes that enabled the UAW's unionization efforts
at a Volkswagen plant in Chattanooga, Tennessee, to overcome defeat in
2014 and 2019 and succeed in 2024. It analyzes the decentralization of
organizational leadership that built increased trust, learning, and
innovation among workers"— Provided by publisher.
Identifiers: LCCN 2025052262 (print) | LCCN 2025052263 (ebook) | ISBN
9781439926406 cloth | ISBN 9781439926413 paperback | ISBN 9781439926420
pdf
Subjects: LCSH: International Union, United Automobile, Aerospace, and
Agricultural Implement Workers of America. Local 42 (Chattanooga,
Tenn.)—History | Labor
unions—Recognition—Tennessee—Chattanooga—Elections | Automobile
industry workers—Labor unions—Organizing—Tennessee—Chattanooga |
Collective bargaining—Automobile industry—Tennessee—Chattanooga |
Labor unions—Organizing—Tennessee—Chattanooga
Classification: LCC HD6515.A82 W35 2026 (print) | LCC HD6515.A82 (ebook)
LC record available at https://lccn.loc.gov/2025052262
LC ebook record available at https://lccn.loc.gov/2025052263

The manufacturer's authorized representative in the EU for product safety is
Temple University Rome, Via di San Sebastianello, 16, 00187 Rome RM, Italy
(https://rome.temple.edu/).
tempress@temple.edu

♾ The paper used in this publication meets the requirements of the
American National Standard for Information Sciences—Permanence
of Paper for Printed Library Materials, ANSI Z39.48-1992

Printed in the United States of America

9 8 7 6 5 4 3 2 1

CONTENTS

PREFACE

Ever since the trade publication *MotorTrend* began designating a Person of the Year, the prize has been reserved for a select cohort of top executives culled from C-suites and boardrooms. An exercise in self-flattery, the annual ritual was a predictable spectacle that attracted little notice beyond automotive hobbyists and industry watchdogs. Until 2024, the list of honorees read like a who's who of influence peddlers and power brokers at some of the world's wealthiest companies.

Bucking the trend, the 2024 *MotorTrend* Person of the Year was a former electrician whose claim to fame was a nationwide strike that brought those very companies to a standstill. By bestowing the honor on United Auto Workers (UAW) president Shawn Fain, even an unabashed partisan rag and beacon of industry like *MotorTrend* could not escape the fact that something was now different.

This book concerns the UAW's decade-long campaign to establish labor representation at Volkswagen's Chattanooga Assembly Plant. Over ten years, the UAW made two unsuccessful attempts to unionize the factory (in 2014 and 2019) before finally succeeding in 2024.

Much like Fain's personal story, the UAW's trajectory from 2014 to 2024 has a cinematic arc. After two successive defeats, the UAW elected new leadership, mounted a successful strike against Detroit's "Big Three,"[1] and remade itself as a fighting union. Volkswagen (VW) had been the linchpin of the UAW's campaign to organize foreign automakers, and after repeated failures, those efforts had finally paid off. Pundits and commentators were quick to craft a heroic narrative of defeat and triumph, proclaiming a reversal of fortune that signaled a sea change for the labor movement more broadly. Countless profiles spun a tale of redemption, setting up a narrative arc in which years of struggle and disappointment encouraged introspection, leading to gradual changes and culminating in the 2024 victory that marked the UAW's ultimate triumph—a tale played out on an organizational scale.

While the dramatic appeal is undeniable, efforts to cast the UAW's victory as transformative or groundbreaking may be premature. Superlatives make for good headlines, and headlines sell magazines, but regrettably, our story is somewhat more complex. First, the UAW's transformation was sudden, unexpected, and fundamental, signaling less the incremental changes associated with learning than a repudiation of the past. However, while the UAW's victory was indeed dramatic, it was also highly contingent, benefiting from a host of contextual factors, some of which were quite distinct from the UAW's internal reform movement. Moreover, at the time of publication, the UAW's gains remain tenuous and beset by vulnerabilities that will only intensify in the future. As of fall 2025, the union's ability to maintain and extend its newfound power remains indeterminate and untested. Thus, rather than maintaining a steady progression toward a triumphant closure, the book draws elliptically toward an open ending, signaling the uncertainty to come.

The central issue animating this book is the problem of solidarity. Though *solidarity* has long been the watchword of the labor movement, it is undertheorized and misunderstood. With the UAW as an empirical lens, I critically interrogate solidarity's basis, formative process, and durability across the three VW election cycles. A conceptual and practical understanding of solidarity is key to an examination of the factors that contributed to the UAW's defeats in 2014 and 2019 and to a sober analysis of 2024 that evaluates the win with appropriate caution, shorn of the hyperbole that pervaded much early analysis.

Solidarity requires the construction of mutual bonds between parties who may have no natural inclination toward fellow feeling. This book argues that genuine solidarity cannot be fabricated or artificially imposed. It cannot be mass manufactured, as one might assemble a car from tens of thousands of components, each fitting together in a preordained and predictable manner. Nor is it an end goal to be achieved or realized through, for example, a union election. Instead, it is an ephemeral state of becoming, destroyed as easily as it is built.

There is a tendency to approach solidarity as one might a construction site. Indeed, it is perhaps no linguistic accident that Jane McAleavy's preferred measure of solidarity, the "structure test," derives from a civil engineering metaphor.[2] In this telling, solidarity must be tested to ensure it can withstand the forces that seek to destroy it, just as one might evaluate the structural integrity of a roadway. Under this logic, solidarity is appraised on the basis of established metrics that are determined in advance: Does it conform to the blueprint? Is it made of the right components? Does it resemble the other solidarities that have already been built?

But solidarity cannot be built *on spec*. Each instance is a custom job uniquely suited to the specific situation. This book exposes the perils of prefab solidarity, not least because *as-built* diagrams showing final specifications often differ sharply from plans. This discussion will be resumed, but for now, I defer the question of solidarity to allow for an exploration of our empirical study.

Acknowledgments

I am grateful to the following people for their contributions to this monograph.

At Temple University Press, I wish to thank my editor, Ryan Mulligan, for his confidence in this project and willingness to see it through to completion. I also appreciate Gary Kramer's ongoing efforts to help this book reach its audience. Five anonymous reviewers offered critical suggestions at various stages of this process. The care and attention with which they approached their task pushed me to refine my argument and improved the final product immeasurably.

Stanley Aronowitz's incisive writing remains my lodestar, shaping the perspective that informs much of this book (despite our occasional disagreements). Until something better comes along, *False Promises* is still the best guide to the challenges and contradictions of the American labor movement.

Mike Elk of Payday Report opened his home, arranged meetings, and helped with logistics in Chattanooga during the early phase of my research. Michael Gilliand's work at CALEB has been a consistent inspiration over the last ten years. Tabitha Arnold's elegant tapestry, which appears as the cover art, is the visual complement to this book's story.

My dissertation chair, Ruth Milkman of the CUNY School for Labor and Urban Studies, saw the potential of this project and pushed me to expand my ambitions. Along with Charlie Post and Stephanie Luce, she provided painstaking feedback on key sections of my dissertation that would later provide the context for Chapters 3 and 4 of this monograph.

Carsten Hübner of the Transatlantic Labor Institute provided key insights and facilitated interviews. John Torpey arranged and coordinated the 2017 Transatlantic Summer Workshop at Ruhr University Bochum, facilitating key overseas connections. Ulrich Jurgens at WZB Berlin Social Science Center, Michael Fichter at Free University Berlin, Ludger Pries at Ruhr University Bochum, Lowell Turner at Cornell ILR, and Daniel Cornfield at Vanderbilt saw the value of this project and offered support at key junctures. I also wish to thank the participants of the ILR Paper Development Workshop in Amsterdam, especially Markus Helfen of the University of Labour in Frankfurt, for their critical comments on what would become Chapter 4. Multiple awards from the PSC-CUNY Adjunct Professional Development Fund and UUP Individual Development Fund helped finance research-related travel.

My colleagues at the Department of Sociology and Interdisciplinary Studies took an immediate interest in my scholarship and recognized the importance of my research. I wish to especially thank my department chair, Hideki Morooka, who has been understanding and accommodating of my writing schedule.

My parents, Connie and Jon, have been a consistent source of encouragement throughout this process.

Finally, I wish to thank the current and former workers who cannot be named but generously shared their stories and time, without which this book would not be possible. Any errors or omissions are mine alone.

The following journals have graciously provided permission to rework and adapt my own previously published material: Chapter 3 incorporates elements from "Unionization at Volkswagen in Chattanooga: A Postmortem," first appearing in *Labor Studies Journal*.[3] Chapter 6 is derived in part from "Rank-and-File Revolt: Insurgency, Power and Democracy in the UAW, 2019–25," which was published in *Labor History*.[4] Chapter 7 draws on portions of "Third Time's the Charm: Assessing the UAW's Decade-Long Struggle for a Union at Volkswagen," from *New Labor Forum*.[5]

ABBREVIATIONS

AC	Administration Caucus
CIO	Congress of Industrial Organizations
COE	Community Organization Engagement
COLA	cost of living adjustment
CTFP	*Chattanooga Times Free Press*
EV	electric vehicle
GFA	global framework agreement
GUF	global union federation
GWC	(Volkswagen Group) Global Works Council
ICE	internal combustion engine (vehicle)
IGM	IG Metall
ILO	International Labor Organization
IMF	International Metalworkers' Federation
MFD	Miners for Democracy
NAFTA	North American Free Trade Agreement
NGLS	new global labor studies
NLRA	National Labor Relations Act
NLRB	National Labor Relations Board
OEM	original equipment manufacturer
PRA	power-resource analysis
SEIU	Service Employees International Union
TNC	transnational corporation
UAW	United Auto Workers
UAWD	Unite All Workers for Democracy

REASSEMBLING THE UAW

INTRODUCTION

Volkswagen's Chattanooga Assembly Plant opened in 2012 to considerable fanfare and immediately attracted attention from the United Auto Workers (UAW), one of the largest unions in North America. As a company with strong works councils and fairly good relations with unions in its native Germany, Volkswagen (VW) was considered an easier organizing target than most other foreign automakers in the United States. While the plant was modest in size by the standards of auto production sites, it had both symbolic and strategic value for the broader union movement. From the beginning, UAW president Bob King was eager to present the VW unionization effort as the flagship campaign of the UAW's southern organizing push. And the UAW was itself the frontline unit of the "Southern Strategy" of the American Federation of Labor and Congress of Industrial Organizations (AFL-CIO).[1] The stakes would be high, for the labor movement's future in the American South was tied closely to the fate of the VW plant.[2] Though foreign auto plants employed over one hundred thousand workers in the South in 2010, the UAW counted none of their workers as members. Therefore, the real possibility of a victory in the South would have represented, for many observers, not only a historic accomplishment but a bulwark against the tide of history. An anticipated victory in Chattanooga was widely seen as a buffer against further decline, if not the opening foray in a broader strategy to organize other transplants across the South.

When it first filed for a union election in 2013, pundits handicapping the UAW's prospects were bullish. Their optimism was not unfounded. Though the UAW had been repeatedly stymied in its previous efforts to organize "trans-plants," its outlook seemed much better this time. VW differed from other automakers in that it had strong union-affiliated works councils in nearly all of its sixty-one plants, and there was reason to believe senior executives desired the same in Tennessee. Moreover, while most foreign companies operating in the United States have treated anti-unionism as dogma, VW had been uncharacteristically cooperative. Breaking with the dominant pattern, the company allowed union representatives to campaign freely in its factory, met regularly with union leaders, and maintained an official position of neutrality concerning the election. The two top VW officials at the plant even wrote a letter to workers speaking positively of the UAW. If there was a foreign automaker that could be organized, if labor was to gain a toehold in the South, if the UAW was to reaffirm its relevance on an (inter)national stage—this was its moment.

But as history now shows, for reasons both endogenous and exogenous, the UAW would need ten years and two more tries to win in Chattanooga. This book explains why. A systematic analysis of strategic challenges and tactical shifts reveals the patterns that persisted across the three elections while also highlighting the distinctions that mark their differences. While taking into account the overdetermined nature of any union election outcome, this analysis shows how the twice-defeated UAW converted its disappointments into success.

Theoretical Frame

Strategic Capacity: First-Order Dimensions

Experts have identified several factors that appear to enhance unionization success rates, even in the face of difficult odds. Borrowing Marshall Ganz's analytical framework[3], I argue that knowledge, motivation, and heuristic processes are essential conditions that facilitate success in an organizing campaign, together creating what he terms strategic capacity, as shown in Table I.1.

First, *knowledge* is a measure of information availability and preparedness—including but not limited to strategic research. Detailed familiarity with their target allows unions to exploit potential weaknesses and anticipate tactical shifts. *Motivation* refers to "the ability to capitalize on opportunities by turning the resources one has into the power one needs."[4] Arguing against resource-mobilization theory, Ganz claims that even resource-rich organizations may fail when said resources are not deployed effectively. Finally, *heuristic processes* involve tactical flexibility and resiliency—the ability to learn from one's mistakes in real time, coupled with a willingness to take risks and seize opportunities as they emerge. Yet, complicating matters, these elements

TABLE I.1. FIRST-ORDER DIMENSIONS OF STRATEGIC CAPACITY	
Knowledge	Institutional memory Pattern recognition Access to salient and relevant information
Motivation	Persistence Commitment Task-oriented focus Effective deployment of available resources
Learning	Resiliency Tactical flexibility Adaptive response to challenges
Innovation	Creativity Risk-taking behavior Openness to alternative interpretations

are overlapping and co-constitutive: for example, encyclopedic knowledge can inspire motivation, just as agility and inventiveness may boost preparedness. Important for my purposes, these components have an underlying temporal logic: access to information is backward looking, drawing on prior experience and learned expertise; motivation rests on an organization's ability to deploy resources in the present, while heuristic processes are future oriented, demanding that a union dynamically adapt to changing conditions and anticipate its opponent's next moves.[5] The unsuccessful 2014 and 2019 campaigns fell short on all three measures, reflecting deficiencies in hindsight, competence, and foresight.

My conceptual map differs from Ganz's in one important respect. For Ganz, "heuristic processes" are a multidimensional construct incorporating a range of indicators across multiple categories. Broadly, these can be divided into two groups: *learning* (gleaning lessons from the past to avoid repeating previous mistakes) and *innovation* (responding on the fly to unexpected challenges through creative solutions that depart—sometimes radically—from established patterns and habits). Though Ganz argues persuasively that learning and innovation are interrelated, the literature is divided on this count.[6] Learning without innovation may lead to interminable cycles of incremental change that do not break free from the past so much as assimilate prior errors through trial-and-error course corrections. Likewise, innovation without learning might generate wildly speculative solutions without practical basis or historical context. We find examples of both in the chapters that follow. Although assessing the interrelatedness of learning and innovation is well beyond this project's scope, I have elected to distinguish them for analytical purposes. Thus, instead of melding learning with innovation under the category of heuristic processes, I offer a gentle corrective to Ganz, treating learning and innovation as functionally distinct.

Strategic Capacity: Second-Order Dimensions

Taken at face value, strategic capacity might seem aligned with the "strategic choice" school of managerial thought, which posits that organizational outcomes stem from leaders' discretionary choices.[7] But here, Ganz introduces another layer of complexity. Though individual leaders may exercise considerable discretion, their decisions are often constrained by deep-seated structures that delimit the range of possibilities. Superimposed on his three (now four) dimensions of strategic capacity, Ganz proposes two additional indicators: *leadership* qualities and *organization* characteristics.[8] As second-order components of his strategic capacity framework, leadership and organization lay the critical groundwork that either enables or inhibits the development of strategic capacity.

Importantly, Ganz treats both leadership and organization not as pure, objective factors but as relational and socially determined. His concept of leadership is less interested in describing the personal talents of individual leaders than in uncovering the contextual and situational factors that allow effective leadership to emerge.[9] Leaders with diverse life experiences and broad networks have a greater capacity to develop effective strategy. Ganz also highlights the importance of internal debate and dissensus within leadership teams, suggesting that leadership benefits from a mix of experienced perspectives and dissenting voices rather than a coterie of yes-men surrounding benighted rulers. To this, we might add a consideration of followership, given that a leader's legitimacy rests heavily on whether followers choose to comply with or resist the leader's directives. In sum, effective leadership cannot be reduced to individual traits alone; it is shaped by the broader relational context.

Similarly, organizations do not exist in a vacuum; rather, they are constituted through a series of inclusions and exclusions. Analyzing potential for strategic capacity requires looking beyond formal bylaws and organizational charts and toward the alliances and "structuring absences" that demarcate the organization vis-à-vis outsiders. Organizations that encourage regular, open deliberation, draw resources from diverse constituencies, and hold leaders accountable are more likely to generate effective strategies. Furthermore, leadership selection processes are key indicators of responsiveness, as they predict whether organizations will adapt to new conditions and seize opportunities. This meta-organizational perspective situates the organization within a broader web of social relations and resource flows.

Focal Moments

As second-order variables, leadership and organization are submerged beneath the more visible first-order dimensions of strategic capacity, only coming into sharp focus during infrequent events that disrupt the smooth functioning of organizational life. To explain this phenomenon, Ganz develops a concept of "*focal moments*," evidently borrowed from social movement scholar John Lofland's "focusing crises."[10] While the weight of inertia normally drives organizations toward stasis and path dependency, these disruptive events shake loose old structures, making space for alternatives to be created in their stead. During such crucial conjunctures, "new kinds of actions are required because some emergent circumstance now demands them."[11]

During focal moments, powerful groups that once seemed part of the permanent landscape find themselves on the defensive, their critics newly emboldened. Such moments refocus the analytical lens, switching emphasis from organizational *behavior* to organizational *composition*. During such inflection points, leaders may find that their grip on power, once taken for granted, is suddenly tenuous. Awash in crisis, questions of leadership style open up to a more fundamental reconsideration of particular leaders' legitimacy. Organizational crises collapse the known coordinates that define organizations, opening opportunities for realignment and renewal.

Though Ganz does not say so explicitly, I contend that innovation is especially attuned to focal moments, arising not from consistent and predictable patterns but from cracks and ruptures. Without an instigating event to set change in motion, organization and leadership take on an apparent permanence that may permit incremental change but precludes meaningful innovation. Focal moments trigger a restructuring of organization and leadership, redounding to the other aspects of strategic capacity (knowledge, learning, motivation). In doing so, they expand the realm of the possible, calling into question that which once seemed beyond reproach.

Central Thesis

Focal moments and the primacy of organizational change are crucial in explaining the UAW's long-awaited 2024 victory in Chattanooga. Repeated failures over two previous election cycles revealed that the union could not win at VW solely through a better organizing approach or tactical improvements—narrowly conceived as "strategic choices." While such granular changes produced marginal gains between the 2014 and 2019 elections, they failed to reverse the fortunes of a union that had become consigned to failure. Injecting the UAW with greater strategic capacity demanded a complete organizational reboot, inviting a reconsideration of the union's essential character. Though the UAW's transformation ultimately stopped short of a decisive break from the past, its thorough reevaluation of organizational priorities spared no sacred cows.

Examining the three election cycles in their totality makes clear that the UAW's defeats in 2014 and 2019 followed organizational and leadership problems that had gone uncorrected. Likewise, the dramatic turnaround realized at the ballot box in 2024 owed less to strategic choices than to the union's ongoing reinvention. Winning in Chattanooga required remaking the UAW on the fly, a process that, as of this writing, is still ongoing and by no means assured.

More generally, this book demonstrates the failure of labor movement renewal through granular reform and reveals the limits of incremental change. The challenges that beset twenty-first-century unionism are intrinsic to the organizational forms and leadership structures at the core of the labor movement's very identity. While American labor leaders rightfully decry the country's hostile legal environment and noxious political climate (mitigated only slightly under President Joe Biden), resurrecting the labor movement will require them to turn their critical gaze inward, abandon habits and practices that no longer generate returns—and even clean house when necessary.

The Limits of Incremental Change

A fine-toothed comparison of outcomes and organizing strategy across the three election cycles reveals that organizational change overshadowed tactical change in growing support for unionization. Table I.2 charts the UAW's strategic capacity over time and also serves as a conceptual map and summary of the text's central argument, further elaborated in the chapters that follow.

In 2014, the UAW performed poorly across all four first-order dimensions of strategic capacity, contributing to its disappointing defeat. A lack of institutional memory and poor dissemination of *knowledge* among unionization proponents meant that the UAW could not adequately prepare for the challenges it would face, as it failed to summon relevant lessons from past efforts to organize transplants. An underresourced and understaffed organizing office constituted a lack of *motivation* under Ganz's framework. The union's *learning* practices were insufficient to adapt to the rapidly changing conditions, leaving it unable to adjust its plans as the situation evolved. While it showed some potential for *innovation*, its bold plans were squandered without a solid foundation.

TABLE I.2. STRATEGIC CAPACITY ASSESSMENT, 2014–2024			
	2014	2019	2024
Knowledge	Moderate	High	High
Motivation	Low	Moderate	High
Learning	Low	High	High
Innovation	Moderate	Low	Moderate/high

Compared to its weak 2014 performance, 2019 looked like a marked improvement. Reeling from its defeat five years earlier, the union increased its knowledge base, mounted a more vigorous campaign effort, boosted its motivation quotient by committing more organizing staff and resources, and developed adaptive learning practices to help it avoid the pitfalls of previous campaigns. Only on the fourth metric, innovation, did the UAW continue to falter. Yet despite faring better on three out of four strategic capacity measures, it lost the election by almost the same margin as it had five years earlier. The comparison between 2014 and 2019 sets up the central question that motivates the book's first half: Why did gains in strategic capacity fail to generate meaningful returns at the ballot box?

The 2024 victory provides a ready answer. Innovation proved the critical factor that distinguished the 2024 campaign from previous efforts—and the secret ingredient necessary for success. All four dimensions of capacity combined for a dramatically revamped organization. Modest improvements are no cure-all, especially when deep structures remain unexamined and unaltered.

The Primacy of Organizational Change

Innovation, the final dimension of the UAW's transformation, does not occur by chance or stroke of luck. Indeed, without prodding, organizations tend to fall into comfortable patterns. Following Ganz, this case demonstrates that innovation is closely associated with second-order dimensions of strategic capacity (leadership and organization), which themselves become salient only during focal moments.

Tables I.3 and I.4 summarize the UAW's leadership style and organizational characteristics at the beginning and end of the study period, which I outline here and cover more systemically in Chapters 4, 6, and 8.

During the 2014 and 2019 election cycles, the UAW's leadership, under the ruling Administration Caucus (AC), adopted a highly centralized and autocratic style, concentrating power in the hands of top-level organizers

TABLE I.3. LEADERSHIP CHANGE		
	2014	2024
Origin	Machine politics	Rank-and-file caucus
Legitimation	Institutional might	Insurgency/contestation
Repertoires	Little salience to constituencies Path dependency Risk aversion	High salience to constituencies Experimentation Calculated risk-taking
Governance	Uncontested or pro forma elections Stifling of caucuses Autocracy	Contested elections Competition between caucuses Diffusion

TABLE I.4. ORGANIZATIONAL CHANGE		
	2014	2024
Deliberation	Delegate based Consensual Univocal	Direct Conflictual Polyvocal
Resource flows	Trickle-down Staff dependent Information rationing	Traversal/circulatory Member-to-member Transparent
Basis of power	Bestowed by corporate headquarters	Immanent to the mode of production
Terrain of struggle	Firm based Managerialist	Industry-wide Sectoral
Relationships	Few in Number Extensive Affiliative	Numerous Intensive Affinitive

and stifling internal debate. AC insiders determined marching orders, often without meaningful input from the rank-and-file members, leading to cronyism and alienating the broader membership. As long as the AC remained in control, the union was unlikely to break free from the cycle of past defeats and resolve the systemic issues that had plagued its previous organizing drives.

The UAW's focal moment came in the form of an embarrassing scandal. Long believing itself all but impervious to challenge, the AC finally pushed its luck too far, finding its top agents in federal court facing multiple felony charges. Though actual criminal behavior was evidently confined to a dozen executive-level officials, the entire organization was tarnished in the court of public opinion. With the union disgraced by scandal, its survival instinct kicked in, and it was forced to address a culture of unaccountability that had allowed such crimes to go unheeded in the first place. For the broader organization, ousting the guilty parties and dressing down their enablers would not be enough. Members were mindful that power has an uncanny ability to reconstitute itself once crises inevitably fade. To survive its existential moment, the UAW would have to reinvent itself on the fly, signaling to its current and future members that it had returned from the brink by making far-reaching changes. In short, the existing power structure was not capable of righting itself. The impetus for change had to come from the outside, in the form of an insurgent movement fully independent of the AC's pervasive influence.

This call was answered by the rank-and-file caucus Unite All Workers for Democracy (UAWD), already strategically positioned to lead the democratic revolt but until then marginalized by a leadership that had spent seventy years consolidating its power. Seizing upon the crisis (with an assist from an assigned federal monitor), UAWD replaced an arcane delegate system with direct elections, opening space for the first genuinely contested election in the union's recent history. Moving swiftly, it toppled the union's existing leadership and installed new representatives in the seat of power. Apart from its takeover of the executive board, UAWD also won key down-ballot victories, preventing the concentration of power that had characterized the previous regime. The UAWD's slate of reform-minded leaders had an approach to organizing that embraced productive confrontations between members, setting in motion broader transformations that reshaped the union's culture from within.

In a testament to the enormity of these changes, within months of UAWD's ascent, the union had embarked on a strike against the Big Three automakers,[12] demonstrating the power of grassroots mobilization and collective action, laying the groundwork for innovative practices, capturing the imagination of the newly mobilized rank-and-file members, and exposing the weakness of the isolated, single-plant approach previously taken by the organization.

From Autocracy to Dispersed Power

The UAW's organizational change finds individual resonance in the 2023 election of Shawn Fain as president of the UAW. His progressive credentials aside, Fain is a somewhat contradictory character. At the level of biography, Fain's story is not a stark break with the past. Like nearly all leaders before him, Fain is a gray-haired white man who worked his way up the UAW's internal hierarchy and eventually became a mid-tier functionary under the ruling AC. By the time of his appointment, Fain was already a seasoned operative, far removed from his early years as a rank-and-file production worker. Moreover, Fain is not a commanding presence in the mold of the classic union firebrand. In public appearances, he leans heavily on a midwestern everyman shtick that borrows less from Mike Quill than Michael Moore. Even his now-legendary appearance at the 2024 Democratic National Convention answered Hulk Hogan's shirt-ripping MAGA reveal with self-deprecating parody.

But one key difference separates Fain from his predecessors. Previous candidates had often run as consensus picks with no serious opposition, shielded from dissident movements by a bureaucracy that insulated leaders from their charges. Thus, though it would be an unfair exaggeration to suggest previous UAW leaders were *merely* a product of machine politics (some presented themselves as reformers), whatever their personal proclivities, they functioned within a system that served to maintain the organizational hegemony of the AC. With Fain, the UAW found a leader forged in the fire of insurgency who would, it hoped, remain indebted to and associated with the contentious process that created him.

From Centralized Control to Deliberation

If history is any guide, UAWD might have been expected to dissolve itself into the union bureaucracy upon seizing power. The antiestablishment passions that drive rank-and-file movements have often proved incompatible with the solemn responsibility of governance. No matter how earnest their political commitments, legal and bureaucratic constraints push leaders toward compromise and can erode deeply held ideals. Though UAWD's long-term viability is unclear, it maintained an active caucus structure independent from officeholders to hold them accountable through 2024.

Previously, operatives complained that the flow of resources and support across the organization remained siloed. Under the new leadership, organizing became a priority at all levels, while Fain encouraged robust debate and discussion among the ranks, allowing diverse voices to contribute to the decision-making processes.

Toward a Recombinant Unionism

Consistent with Ganz's meta-organizational perspective, the UAW's transformation was not confined to the union's internal affairs but also resituated the organization's stance toward external actors, reconstituting its relationships and reorienting its institutional priorities. Under Fain, the UAW disentangled itself from alliances and agreements that had proved disadvantageous, while building out new coalitions and connections.

From Unionism in One Company to Sectoral Bargaining

In 2014, the UAW's organizing approach treated VW as a special case, distinct and largely disconnected from prior campaigns at transplants and other ongoing work. In doing so, it kowtowed to the company, positioning VW corporate headquarters as the site of power toward which union leaders must genuflect. Subscribing to corporatist logic, the UAW came to believe that power must be summoned from on high and deferred to the company to facilitate unionization.

Workers felt alienated from such alliances, as they operated at a level removed from workers' immediate, tangible challenges. Broad transnational agreements failed to engage the complexity and specificity of local struggles. Existing global governance tools increased the distance between the shopfloor and corporate headquarters, undermining union democracy in the process.

Effective transnational solidarity remains a challenging proposition. Much as I reject the idea that the solution to autocratic leadership styles lies in a pure and naive rank and filism, I am similarly skeptical that the answer to the problems of global governance lies in the romance of bottom-up localism. Instead, unionism across borders requires eliding the neat binary between global and local. The connections that underpin economic interdependence can be manipulated, either stretched or compressed, to reposition relevant actors in the social field. Effective transnational solidarity is less about bridging geographical distances and more about creating numerous, meaningful, and intimate connections between workforces. Rather than aiming to tame the boundless power of transnationals by pegging them to known coordinates, labor might take the supposed "placelessness" of global capital at its word, matching the flightiness of the transnationals with an agility of its own.

The focal moment of the 2023 scandals provoked the UAW to adopt just such a dynamic strategy in 2024. Existing arrangements of social actors gave way to the proliferation of connections, replacing unidirectional global governance models with the multidirectional, circulatory flows of resources. Against the rigidity of traditional spatial distinctions, labor transnationalism requires pushing beyond territorial limits, coupling the extensive reach of global capital with intensive solidaristic practices.

The firm-centric approach of 2014 gave way to pseudosectoral bargaining as the UAW began to connect its organizing targets more explicitly, breaking down the strategic firewalls that had previously isolated different campaigns. Pitting firms against one another to generate competitive pressure, the UAW sought to leverage its existing power to influence organizing efforts among transplants, recognizing that a good contract with major automakers at the Big Three could serve as powerful advertising for the union. This sectoral strategy not only reflected industry conditions but also rhymed with the perspective of workers, who were quick to draw parallels between competing firms, forging lateral alliances that emerged organically.

Meanwhile, the union replaced hierarchical and formulaic organizing models with new experimental methods, drawing on what Mark and Paul Engler term "momentum-based organizing."[13] Under this model, successful organizing efforts lead to the viral spread of copycat actions, creating a sense of movement through contagious waves of action. The Big Three strike's success had already demonstrated the power of collective action and provided a template for future organizing efforts, setting the stage for the UAW's return to Chattanooga with a fresh mandate. But the strike's influence also inspired a comprehensive strategy targeting all nonunion automakers simultaneously, rather than isolated campaigns confronting them individually. Union staff now relinquished control over the day-to-day work of organizing, allowing members to take a more active role, while training and developing organic leaders from within the membership.

By fostering lateral connections among workers, between firms, and across boundaries, momentum-based organizing spreads virally, traversing the lines that delineate conventional organizing drives. Connections, once few in number, were allowed to proliferate. Affinitive ties replaced affiliative ties. The once-risk-averse organization turned toward experimentation. In its best moments, this recombinant unionism is immanent to the mode of production, indifferent to the often arbitrary distinctions that fragment workers' collective power.

Prefigurative Practices

In the final analysis, the UAW's strategic shifts over the decade in question reflected its transformed organizational character. The strategic advances previously outlined were possible not merely because Fain and his slate ran as "change agents" but because their leadership styles prefigured the practices the UAW now embraced. The UAW failed in 2014 because it privileged *extensive* alliances over *intensive* connections, institutional might over contentious struggles over power, and the certainty of fixed habits over experimentation. The 2014 UAW developed a strategy that centered the firm's perspective, rested on the power of distant and aloof institutional bodies, had little relevance to the U.S. context, and depended on the company's good graces to hold itself accountable to a global governance pact. In short, the UAW's 2014 campaign mirrored the union's anti-democratic character. After its "focal moment," the newly innervated UAW could no longer abide the continual recycling of failed organizing strategies lest it betray its own ideals. By 2024, it had sidelined the works council and dismantled the global governance agreement, creating space for the rank and file to step into the fray and make the campaign their own. By 2024, its tactical repertoire took its inspiration from the rebel caucus that was remaking the union from within, and the new organization radically expanded its capacity, embracing the resources of the broader membership.

Power Resources

In adopting Ganz's strategic capacities frame, this book rejects a more widely accepted power-resource analysis (PRA) model.[14] PRA suffers from a reified concept of power, viewing it as an object to be possessed rather than a potential to be deployed. Under PRA, power is accumulated and held in reserve, much like one might stockpile commodities. But reality often departs from the idealized, laboratory-like conditions in which such grand theories flourish. When PRA hits the ground, the precise mix of resources will differ dramatically across unions and contexts. Relatedly, PRA largely ignores the processual dimension of power. Given historical contingencies, differing organizational capacities, and shifting sociopolitical dynamics, unions may struggle to deploy all power variants equally. Indeed, new research has begun to move toward modeling the various dimensions of power as components of a master index,[15] invoking the possibility that power sources might be interchangeable. Finally, PRA does not consider organizational change, demoting innovation to a second-order effect.

Without denying the importance of PRA or the impact of legacy effects that interdict the introduction of new theoretical frames, I believe strategic capacity may be more illuminating. Against PRA, strategic capacity is an agential model that shifts the emphasis away from the organization as a static entity "having" varying degrees of power and toward the activated work of organizing, in which the exercise of power is its realization. Furthermore, unlike PRA, strategic capacity is iterative and time sensitive, attuned to the past, present, and future. Finally, strategic capacity stresses the processual aspects of power, emphasizing how organizations identify, understand, and mobilize the various forms of power to which they have access.

Data and Methods

There has been no shortage of journalism and academic writing on the VW elections.[16] However, existing research has often failed to incorporate rank-and-file perspectives.

Data collection consisted primarily of semistructured, long-form interviews with workers across the ideological spectrum, totaling over two hundred hours of audio transcripts. To preserve anonymity, I have used pseudonyms for all workers quoted in this book. I have identified public figures by name when possible, although some were unwilling to speak on record unless anonymized. Where appropriate, I have supplemented the interview data with my own interpretive findings.

I also gained access to a Facebook group used by UAW leaders for internal coordination, which I quote directly where relevant. This repository is the most extensive written account of the unfolding of the UAW's campaign as seen by member organizers. Once dismissed as a mere sideshow or, more generously, a supplement to traditional face-to-face organizing endeavors, social media has proved to be fertile terrain for labor organizing. The Facebook data are invaluable in providing a real-time snapshot of members' attitudes toward the union, the organizers, local management, and assorted German interlocutors. They offer a running commentary on the campaign, unobstructed by the self-censorship that often contaminates more public forums. Among other revelations, the Facebook group provides a record of the spirit of solidarity that the drive initially cultivated and its rapid disintegration in the aftermath of the 2014 defeat. Equally important, the Facebook data provide a lasting record of internal disagreements and disputes among worker-organizers.

Methodologically, this study resembles a natural experiment, comparing three union drives launched in the same setting with distinctly different strategies. But conditions changed over the decade, undermining the quasi-control setup and introducing error through selection-maturation interaction. Internal validity is, therefore, a major challenge in a study of this type. Though this book presents itself as a comparative study of election outcomes across three elections, VW's workforce is statistically incommensurate over a ten-year period. Of the workers employed at the plant in 2024, only a fraction worked for the company in 2014, and still fewer were intimately involved in the organizing campaign at the time. Though scholars disagree on the maximum acceptable attrition rate for longitudinal analysis, attrition rates exceeding 50 percent are generally thought to seriously compromise findings.[17] To put it bluntly, this study falls short of meeting the standard.

Moreover, the fact that attrition was almost certainly nonrandom further biases the results. Nonetheless, because snowball sampling is already nonrandom, and since sampling procedures varied somewhat across the three elections, such objections may be moot. Further, as is often the case with "natural" experiments, circumstances differed so radically across the three elections that it may be impossible to isolate causal factors. Nonetheless, after cross-checking and validating my interview data against contemporaneous reports, I have identified qualitative shifts in attitudes and behaviors that explain the differing election results.

A final methodological point concerns election outcomes. Union elections are all-or-nothing affairs; there is no consolation prize for almost victors. In 2014 and 2019, the union finished within three percentage points of victory.[18] However, because hundreds of workers abstained from voting, the proportion voting no as a subset of the self-selected "sample" (actual voters as a percentage of eligible voters) is perilously close to the allowable margin of error at the 95 percent confidence interval. Though there is no reason to suspect that union supporters were overrepresented among abstaining voters, we cannot exclude this possibility on its face. Put differently, there is a slim but nonzero chance that the scale might have tipped in the union's favor if all workers had voted (or if a different but equally large group had abstained). Obviously, from the union's perspective, coming close did not lessen the blow of defeat, but for all the criticism the UAW endured, it never missed its mark completely. In retrospect, its two-time near majority provided a foundation that sustained the campaign over the course of ten years. On the other hand, in 2024, the margin of victory was so decisive that full voter turnout could not have affected the outcome. The union's 73 percent win in 2014 cannot be dismissed as a statistical anomaly.

Structure of the Book

This book's core empirical chapters provide the central narrative framework by tracing the development of strategic capacity across the 2014, 2019, and 2024 election cycles. Chapters 3, 5, and 7 are devoted to what I have termed first-order dimensions, while Chapters 4 and 8 pull back the curtain to examine second-order dimensions, showing how leaders and their organizations affected strategic capacity.

Within the broader structure of the book, Chapters 4 and 8 function as a paired dyad, examining the UAW from a meta-organizational perspective and showing how its organizing strategy in Chattanooga was a by-product of its relationships with other actors and its positioning within broader systems of global governance. Within this framework, Chapter 6 serves as the linchpin on which the narrative turns, outlining the 2023 focal moment that ushered in the union's reinvention. The balance of the book completes our understanding of these events with relevant background, commentary, and theoretical analysis.

We begin with Chapter 1 ("Solidarity Divided"), which provides a high-level overview and crucial scene setting by situating Chattanooga, VW, and the UAW in their historical and economic context, and positions the firm and the union against the shifting relations of automotive production. The chapter discusses how macroeconomic trends have increased fragmentation and deepened divisions among autoworkers, contributing to a decline in union strength. Key factors such as capital flight, corporate capture, and subcontracting or outsourcing are identified as significant destabilizing factors. Despite these challenges, the chapter suggests this restructuring may open new pathways to solidarity, even as other opportunities are foreclosed.

Chapter 2 ("Solidarity Denied") describes the union's eight organizing challenges in 2014, some of which would recur in 2019 and 2024, and each of which held valuable lessons the union would draw upon over the decade. Table I.5 summarizes the real and perceived obstacles that complicated the UAW's organizing drive over three election cycles, first introduced in Chapter 2 and readdressed in subsequent chapters.

Chapter 3 ("Solidarity Subdued") explains how the union responded to the obstacles presented in Chapter 2. Four analytical foci emerge, adapted from Ganz: knowledge, motivation, learning, and innovation. In 2014, the UAW fared poorly on all of them. Briefly, it should have known what to expect;

TABLE I.5. OBSTACLES TO UNIONIZATION, 2014–2024								
	Honeymoon effect	Union substitu-tion	Promises and improve-ments	Delay	Political interfer-ence	Disinvestment threat	Union suppres-sion	Deflection and dis-tancing
2014	X	X	X	X	X	X	X	X
2019		(X)	(X)	(X)	X	(X)	X	
2024		(X)			(X)		(X)	
Key: X = major impact; (X) = minor impact; [Blank] = no impact / does not apply								

it was unprepared for the fight of its life, and it did not adapt quickly or easily to changing circumstances. The chapter sets a baseline marker for understanding how these lessons were applied in the 2019 and 2024 organizing efforts.

Chapter 4 ("Solidarity's Enclosures") returns to the 2014 election with a wide-angle lens, showing how problems of leadership and organization inhibited the development of strategic capacity and contributed to the unforced errors noted in Chapter 3. Centering the transnational dimensions of the 2014 election, I show how the UAW's global-level alliances came to reflect the union's antidemocratic and bureaucratic character. Appealing to corporate headquarters and prioritizing high-level, broad agreements left the union poorly equipped to address the specific challenges faced by workers in Chattanooga.

In Chapter 5 ("Solidarity Deferred"), I turn to the 2019 election, which was heavily informed by the defeat five years prior. As Table I.5 shows, some of the most serious obstacles from 2014 had become non-factors by 2019, and others had diminished in importance. Moreover, the union scored high on three of four strategic capacity criteria. Notably, it gleaned important lessons from its 2014 defeat and took appropriate measures to avoid repeating them. Yet despite fewer obstacles and a more sophisticated organizational response, the UAW could not move the needle. Even after improving its knowledge, motivation, and learning, a point deduction for weak innovation proved debilitating. I infer that innovation may be an indispensable precondition—a sine qua non causal factor without which strategic capacity is unattainable.

The emphasis on innovation sets up a discussion of the literature on organizational failure and change, which is taken up at the beginning of Chapter 6 ("Solidarity in Conflict")[19] and further developed in the remainder of the book. Chapter 6 serves as our narrative's inflection point, describing a series of events that, while not directly affecting the election in Chattanooga, shifted the terrain of struggle. Specifically, in 2023, an insurgent movement toppled the UAW's reigning leadership and installed a maverick slate of newcomers with progressive political bona fides and a more confrontational approach to organizing. In the process, the union's culture was reshaped from within, resulting in democratic reforms and energizing the rank and file. I recast the classic question of union democracy in terms of contestation and insurgency, emphasizing the contingent character of any democratic reform. These events provided the UAW with a crucial reset, allowing it to return to Chattanooga in 2024 with a fresh mandate free from the dragging weight of past defeats.

Chapter 7 ("Solidarity Reinvented?") applies our strategic capacity framework to the 2024 election. As Table I.4 shows, barriers to unionization declined significantly from 2019 to 2024, while shifts in the political landscape, changes to labor laws, and differences in the demographic composition of the workforce all benefited the UAW. Yet while counterfactual historical claims are notoriously problematic, the organizational renewal discussed in Chapter 6 ultimately proved decisive. Better knowledge, motivation, and learning may have helped avoid a repeat of 2014, but only innovation allowed the UAW to move beyond the setback of 2019 and secure its desired outcome in 2024.

In Chapter 8 ("Solidarity's Intimacies"), I show that the union had to break with the past to reinvent itself. The UAW's transformation involved realignment across spatial, relational, and processual axes, repositioning relevant actors in the social field. Leadership and organizational changes reshuffled alliances, creating a basis for new modes of struggle while breaking old connections and forming new ones.

The Conclusion ("Solidaristic Futures?") reunites the discussion of space and relationality from Chapters 4 and 8 with the strategic capacity focus of Chapters 3, 5, and 7, all while returning to our original question of solidarity, now with an orientation toward the future. The geographical and logistical changes brought about by the rise of electric vehicles (EVs) are likely to further disrupt global production networks. Nonetheless, the UAW now views the EV sector as an opportunity to extend its organizational reach, rather than a threat to its traditional base. Organizations achieve solidarity not by dredging up history or rummaging through a familiar tool chest but by transcending the past with an eye toward the future.

1

SOLIDARITY DIVIDED

The Dispersal and Fragmentation of the Auto Industry

T
his chapter introduces our research site by positioning it against the
context of the changing geography and organization of American
automobile production at the beginning of the twenty-first century.
Though the auto industry's union density was once unparalleled, by the time
of the study period, a series of macroeconomic trends had increased fragmentation and deepened divisions between autoworkers, contributing to union
decline. Specifically, I highlight capital flight (strategic decoupling), corporate capture (strategic recoupling), subcontracting and outsourcing (vertical
disintegration), and insourcing (vertical reintegration), drawing on VW-spe-
cific examples throughout. In the process, we are introduced to key members
of the local and state political establishment, including former U.S. senator
Bob Corker, former Chattanooga mayor Andy Berke, former governor Bill
Haslam, and current governor Bill Lee, as well as major figures in the state and
local economic development agencies.

Alongside this story, we witness the rise and fall of the UAW, which came
to power during the advent of wall-to-wall industrial-style unionism, only to
watch its power decline amid the aforementioned macroeconomic pressures,
ultimately accepting "tiered" and "nonstandard" employment arrangements
in a bid to keep the Big Three afloat. The UAW is a storied union with long-
standing ties to the Big Three in the mid-Atlantic and Great Lakes regions, but
it has had little to no success organizing foreign automakers in the United
States. The combination of factors described in this chapter all contributed
to the UAW's structural decline, and it has yet to fully take advantage of emer-

gent countertendencies, such as insourcing. The reorganization of production and the rise of nonstandard employment relations also served as significant barriers to organizing in Chattanooga.

This heady combination of internal and external factors contributed to the dramatic weakening of the UAW by the time it set its sights on Chattanooga. As a former Congress of Industrial Organizations (CIO) manufacturing union whose membership had hemorrhaged since its midcentury peak, the UAW was the icon of postindustrial union decline. Auto work was once credited with raising living standards among the broader working class in the postwar era, but that promise has faded for new hires and younger workers, even at legacy auto plants today. At the time of the UAW's initial push to organize VW, its image had been battered amid two rounds of concessionary bargaining. While it had experienced high-profile success organizing among higher education workers and others beyond its original jurisdiction, it struggled to retain and organize its traditional base. Its membership rolls numbered 386,677 in 2010, down 75 percent from its peak of 1.5 million in 1979, with autoworkers constituting a minority of its constituents.[1]

Nonetheless, though the overall trend suggests a reduction in worker power, the fragmentation and restructuring of the industry have given rise to certain vulnerabilities with potentially destabilizing effects, complicating matters somewhat. If history is any guide, the union movement will adapt to the reorganization of the labor process, discovering new pathways to solidarity even as others are foreclosed, as explored further in this book's Conclusion. Therefore, though the dissolution of the old production system hurt labor's short-term prospects, its long-term consequences remain undecided.

Fragmentation of the Automotive Production Process

Capital Flight (Strategic Decoupling)

The unique features of automobile production have hastened a spatial reorganization of the automotive industry, negating time-honored customs and challenging long-standing assumptions. The unionization of the American auto industry in the 1930s was aided enormously by its geographic concentration, with all companies headquartered in Detroit and up to 80 percent of all production clustered within a two-hundred-mile radius of the city in southeastern Michigan. Because of these congregative effects, Detroit became synonymous with the automobile industry and, by extension, the labor movement.

In the immediate postwar era, the UAW was a blue-chip union whose accession paralleled the rise of American industrial might. It was the paradigmatic exemplar of what has been described as the "postwar social contract"—a tacit agreement between labor, corporations, and successive Democratic administrations to trade militancy for predictable wage increases.[2] This is not to suggest the path was smooth—the UAW leadership often found itself embattled by Black radicals and other segments of the Left, as well as other dissidents.[3] But until the early 1970s, Walter Reuther's brand of dispassionate, center-left corporatism reigned supreme and won him unlikely allies in Washington.[4] Importantly, the UAW was able to maintain and maximize density without new organizing during this era as the only assembly plants in the United States at the time were U.S.-owned and U.S.-operated firms with standing union contracts. Eventually, the industry sought to escape these pressures through what Beverly Silver has termed a "spatial fix": capital flight to areas where labor is weak.[5]

The conventional view of globalization holds that mass production fled the United States to low-cost markets in the Global South as economic restructuring and global competition imperiled homegrown industry during the last three decades of the twentieth century.[6] Yet, particularly in the case of the auto industry, this characterization is misleading, if not wholly inaccurate. The auto industry in the United States as a whole has not shrunk in terms of vehicles produced annually, and employment in all auto-related jobs was actually higher in 2025 than in 1983.[7] Indeed, foreign direct investment by European and East Asian firms has more than compensated for the job losses of the Big Three. Between 1990 and 2007, even as the Big Three closed twenty-nine North American factories, foreign automakers opened twenty-four North American plants, partially offsetting job loss among U.S.-based firms and challenging the narrative of persistent deindustrialization.[8]

The perception that the auto industry is in decline likely stems from the reality that its unionized sector has indeed hemorrhaged, as the Big Three spin-off parts suppliers and foreign-owned nonunionized "transplants" gain market share from unionized legacy plants. By 2022, total union membership in the sector had declined by 73 percent from its postwar peak.[9] This change parallels the industry's geographic shift in the United States from the Midwest to the Southeast, part of a deliberate strategy to reduce labor costs and avoid militant workers, which was crafted in response to economic uncertainty and loss of market share to Japanese imports during the 1960s and 1970s.[10] Even as the Heartland shed auto jobs, seven southeastern states saw the industry grow, and foreign-owned plants now account for more than half of auto production.[11] The shift is even more dramatic in the emergent EV sector, where 66 percent of all planned jobs are now headed for the South, compared to only 26 percent in the Midwest.[12] (All Big Three plants in the South operate under union contracts, but this is due to their inclusion under nationwide bargaining agreements. In contrast, most transplants remain unorganized.)

Nonetheless, we must avoid a too-simple reading of the American auto industry that presents the North as inherently more hospitable to unions than the South. The reality was always more complicated. While much has been made of southerners' inbred hostility to unionization, the geographic specificity of the South has often been exaggerated. New competition from foreign firms served as a check on the aspirations of northern workers, undercutting the UAW's core bargaining power by holding down wages. Meanwhile, since the 1970s, unions in the North had steadily atrophied, undermining their position by settling for weak contracts and succumbing to corruption. They became less appealing to southern workers and preserved their northern stronghold only by dint of legacy effects. Arguably, endogenous factors, rather than exogenous situational or contextual circumstances, were responsible for splitting the industry along regional lines.

Corporate Capture (Strategic Recoupling)

Economic geographers use the concept of *strategic coupling* to describe elaborate courtship rituals between regions pursuing investment and firms seeking to build physical capital. Site selection requires balancing multidimensional considerations, not all of which are reducible to strict cost-benefit analyses. But the South owes its popularity in no small part to the generous payouts it offers potential suitors and its historical aversion to unions. This section documents how Chattanooga's political elites and city officials used monetary incentives and cultural tropes to attract foreign direct investment.

It is easy to dismiss the South, which was late to industrialize and still retains vestiges of its chattel slavery system to this day, as a backwater or, to use the Wallersteinian paradigm, as a "peripheral" zone economically dependent on and subservient to the "core" urban centers of the North. But this image is increasingly outdated. Today, the South's "business-friendly" climate has given rise to America's second Automobile Alley. In the auto industry's late capitalist diasporic form, greenfield construction sites have centered in the South, which now attracts more foreign direct investment than any other part of the country.

While the globalization literature has tended to privilege the global North-South divide, the South now functions as a perverse mirror of the Global South in terms of its right-to-work laws, low wages, and active courting of industry through heavy economic incentives.[13] While northern states are also keen to lure major manufacturers with tailor-made financial packages, they cannot compete with the South's promise of a union-free environment. Indeed, the allure of the South derives from its low rate of unionization—a point that is not lost on industry. While northern autoworkers have proved expensive and difficult to manage, most southern auto workers have little prior experience with unions, and preexisting anti-union sentiment, consistent with local attitudes, runs high. In capital's relentless search for new labor markets, the nonunion South has become a key frontier in the global race to the bottom. German carmakers have described the South as "Europe's Mexico," and promotional materials have pitched the South as a low-wage haven, likening it to a developing country.[14] A report from one of the world's leading business consulting firms compared the South favorably to China, noting its "flexible unions/workers, minimal wage growth, and high worker productivity."[15]

Despite the industry's internal fragmentation and geographic dispersion, it has rapidly consolidated its holdings and strengthened its regional power base. Because vehicles and major components are high in value, heavy, bulky, and difficult to transport, manufacturers tend to collocate their facilities close to the end consumer. Since the 1930s, vehicle plants and their suppliers in the United States have clustered along the I-75 corridor in the original Automobile Alley. This geographic concentration cuts back on logistical overhead, resulting in lower shipping costs for the customer.[16]

Union avoidance is among the primary pull factors. The challenges unions face as they attempt to organize the South are well documented and have deep historical roots.[17] Though pockets of union density once existed in certain industries (e.g., steel) and specific cities (e.g., Birmingham), the South as a whole has historically resisted unionization, particularly in key sectors such as textiles.[18] The failure of union leaders to develop a coherent and appropriately ambitious southern strategy led to the decimation of southern unions as mass industry fled.[19] Toyota, Kia, BMW, and Honda actively pursued sites that seemed impervious to unionization, expressing a clear preference for right-to-work states and locations far from existing unionized operations.[20] But the growth of transplants in the United States was initially a means of circumventing the Ronald Reagan administration's import restrictions, which were designed to shield U.S.-headquartered automakers and American workers from foreign competition. Early on, the UAW, staring down the prospect of capital flight at its legacy shops, backed the transplants, reasoning that the new factories would create opportunities for displaced autoworkers who could then be organized.[21] The UAW underestimated how much these Japanese (and later German) companies would resist its entreaties.

While there are pockets of relatively high unionization, Tennessee is undeniably difficult terrain for union organizing. With 2.8 percent union density in the private sector, it is among the least unionized states in the country, with the vast majority of its union members working in rail, aviation, government, and other industries that are exempt from right-to-work laws.[22] Like all adjacent states in the Midwest, Midsouth, and Deep South, Tennessee is a right-to-work state, where unionization is notoriously difficult.[23] From the perspective of local leaders, substandard pay is not a source of embarrassment but a point of pride. The Chattanooga Regional Manufacturers Association released a brochure boasting *Chattanooga Makes Sense for Manufacturing*, which gleefully noted that the region's "cost-competitive wages are below national norms, including total average industrial earnings (83%) and manufacturing wages (75%)."[24]

Apologists for the regional wage differential often point to contextual factors. Transplants select sites where the opportunity structure is limited and viable alternatives for unskilled, inexperienced workers are scant. The relatively high pay transplants offer is more than enough to entice an eager pool of applicants. From a rational choice perspective, transplants pay just enough to make the union advantage less appealing in light of the attendant risks. Though pay is low by national standards, it is relatively good for the region.

True to form, VW's proposed labor standards in Chattanooga represented a sharp break with the company's "high-road" reputation.[25] Rather than paying wages on par with German or domestic American automakers, it offered a compensation package consistent with that of other Japanese and European transplants in the South.[26]

Bringing VW to Chattanooga did not come cheap. In the 2000s, the state created elaborate incentives to establish an auto-manufacturing megasite in the region, intended to lure both assembly plants and suppliers, as well as the necessary logistical infrastructure to create seamless connections between the system's components. The three German automakers currently operating in the United States each have plants within a two-hour drive along the I-75 corridor and tend to share suppliers. In return for locating the plant in Tennessee, VW received $300 million in incentives, or about $222,059 per job, not including in-kind payments, such as upgrades to water, sewer, and electrical systems, or the construction of a dedicated interstate exit.[27] Only forty-eight companies have received larger payouts from states, and VW's deal ranks fourth among automakers (excluding allocations tied to the 2009 bailout). If public subsidies are calculated on a cost-per-worker basis, VW almost certainly ranks first.[28] By some measures, state and local entities subsidized at least one-third of the company's initial investment.[29]

Corporate capture in Chattanooga was often personality driven, with individual politicians presenting the region's aggressive pro-business agenda in distinctly personal terms. As in many small southern cities, Chattanooga's entrepreneurial class never matured into a fully formed bourgeoisie. Instead, the city's leaders maintain and legitimize their power through barely concealed systems of personal patronage and paternalistic political and economic relationships. With the notable exception of Mayor Andy

Berke, a progressive liberal who broke with convention and offended up-holders of the status quo, Chattanooga has always been governed by land-owners. Chattanooga's developers dictate city policy, not through lobbying but by actively controlling the policymaking apparatus. Senator Bob Cork-er is the perfect exemplar of this system, as a former real estate developer-turned-politician who often described himself as the city's caretaker and benefactor. He couched the struggle against the UAW in deeply personal terms, constantly reminding workers that he had conceived the idea of de-veloping Enterprise Business Park while speaking of his love for the city and region.[30] In a Trumpian flourish, Corker bragged to reporters that he had been solely responsible for VW's recruitment and boasted that he had gotten to know former CEO Martin Winterkorn "really, really well."[31] Corker's stated contempt for the UAW may have been linked to a certain variety of southern paternalism. A German observer presciently described southern-ers as lacking in self-confidence and preferring the security of a self-pro-fessed caretaker—be it VW or Corker—to the uncertainty of a self-directed future. Corker went on to play a key role in VW recruitment and emerged as a leading voice of the opposition movement in 2014.

Following Corker's lead, even purportedly neutral city leaders served as gracious hosts, hoping to guarantee a return on their investment. City lead-ers readily acknowledged that Chattanooga's low unionization rate was in-strumental in their efforts to court industry. And the city was more than willing to accommodate VW's demands. During the site selection process, when VW executives complained that the proposed business park was too unkempt, city officials invited a small army of earthmovers to clear and raze the site, staging photo ops in the process.[32]

On the other hand, as with several elements of our story, heroes and vil-lains were often replaced by more complex characters. For example, Chatta-nooga's mayor Andrew Berke came from a family of leftist attorneys—some of whom practiced employee-side labor law—and flouted his progressive credentials. Berke stood out among local leaders for his agnosticism with respect to the vote. Unlike Haslam and Corker, both business owners and property managers, Berke was not landed gentry, and his family law practice often took on pro bono cases to help low-income clients. While economic development figured prominently in his administration, he had less stake in the vote outcome than the landowner class. Berke adopted a position of studied neutrality, refraining from taking any formal position on the drive while reiterating his commitment to job creation. While the partisan battles played out, he said he would stay focused on helping VW succeed and expand, irrespective of how the union election was decided.

Moreover, Chattanooga's status as a Democrat-controlled city seeking to boost its urbane credentials and lure creative millennials tempered some of the more overtly pro-business attitudes on display in outlying rural areas. The city's quasi-progressive self-presentation stands in stark contrast to the dominant norm in the South. In this sense, the arrival of a company with a reputation for top-flight engineering meshed nicely with Chattanooga's image of itself as a progressive, tech-friendly city. In the 2000s, Chattanooga had semisuccessfully branded itself as a tech hub, recruiting a series of marquee brands, including OpenTable and SurveyMonkey (both of which have since fled to greener pastures). The downtown corridor attracted bike-share programs, coffee shops, street murals, and other signposts of the creative class. Chattanooga garnered accolades for sustainability and self-consciously positioned itself as a competitor to progressive southern enclaves like Asheville, North Carolina, and Athens, Georgia. VW continues to play a prominent role in the city's marketing pitch as it works to lure other potential investors. Even beyond supply chain effects, VW helped establish Chattanooga as a boomtown, drawing entrepreneurial capital in the form of boutiques, coffee shops, and auto component makers.

Chattanooga's ongoing cultural rebrand also helped align VW with its host city. The city's steel industry and mountainous terrain had once earned it the moniker "Pittsburgh of the South." But these same mountains trapped industrial smog, and Chattanooga frequently ranked among the most polluted cities in the country. Through the 1980s and 1990s, the city suffered from a rash of deindustrialization and depopulation as its industrial base began to disintegrate. It was not until 2011 that Chattanooga's population, buoyed by VW's arrival, returned to the same level as in 1980. By then, the city had begun to rebound from its late twentieth-century nadir, but its resurgence depended on the seasonal tourism industry and rested mainly on outside investment in its downtown commercial district, centered on a riverfront aquarium. While they drew considerable attention, these redevelopment initiatives amounted to little more than window dressing. Sustainable, living-wage jobs were largely missing from Chattanooga's self-presentation as a city on the rebound, making VW's long-term investment in a relatively well-paid workforce all the more vital.

But, as with Baltimore's Inner Harbor and Boston's Faneuil Hall, redevelopment created a self-contained enclave with little impact beyond its boundaries. While these late twentieth-century urban redevelopment initiatives concealed their class politics more effectively than the urban renewal projects of the 1960s, the resulting progress was limited in scope and scale, failing to significantly benefit Chattanooga's residential neighborhoods. The city's development pattern has historically been centered on downtown, a geographically distinct, densely populated square mile by the waterfront and the only portion of the metropolis where land use skews toward urban. In the remainder of the city core, dominated by single-family housing tracts and sprawling strip malls, there are few signs of new investment. The downtown scene is the version of Chattanooga that features in glossy brochures and earns breathless write-ups in the likes of *Outside* magazine's Best Places to Live.[33] But while Chattanooga's downtown boom has brought tourist dollars, job creation remains anemic and largely limited to low-paid positions in the retail sector. Following the Richard Florida model, which has proved successful elsewhere (while attracting its fair share of critics), the city pushed for a downtown Innovation District to lure coveted tech jobs and attract affluent residents who might stabilize the downtown housing market. While VW never made any serious financial commitment to downtown, city officials sought to bridge the gap between their downtown renaissance and suburban business parks by persuading VW to align itself with urban redevelopment. A memorandum of understanding committed Chattanooga to the city's Innovation District, but as of this writing, that "commitment" remains symbolic, as the Innovation District has yet to bear the fruits of VW's involvement.

For now, VW's immediate impact is confined to the suburban region it occupies, a thirty-minute drive across a steep ridgeline from downtown. In contrast to early- and midcentury northern plants, which were typically adjacent to prominent working-class communities, VW Chattanooga has no real spatial relationship to any definable neighborhood or even to the city itself (though it is nominally within city limits). Located in a semirural section of town on the grounds of a former weapons plant, the site was fairly inaccessible until the state Department of Transportation built a dedicated exit off Interstate 75. Although it is technically part of the city, the industrial park and its surroundings lack almost all the distinctive elements that define urbanism as a social phenomenon. If urbanism is defined by chance encounters between strangers and cosmopolitan sophistication,[34] Chattanooga stretches the term's meaning beyond recognition. Near the plant, an outlying region that was largely rural until recently was transformed into a bedroom community of suburban housing tracts. Although the average home price is out of reach for blue-collar workers, managers and executives account for much of this exurb's population.

This section has shown the outsize role of politicians in attracting VW and ingratiating themselves with its executive team. This courtship ritual contributed significantly to VW's selection of Chattanooga, even if the benefits were often one-sided.

Outsourcing (Vertical Disintegration)

In the 1930s, the Ford River Rouge plant famously accepted wrought iron off barges at one end and churned out finished vehicles at the other, manufacturing most parts and components in-house. Even through the 1980s, automakers were known for vertically integrated production networks supplied by a few directly owned subsidiaries, with outside contracts limited to basic commodity items.[35] By the 1990s, this model had disintegrated, with major automakers spinning off their subsidiaries and replacing fully integrated production lines with multitiered supplier networks producing modular components. Today, the very idea of an automobile factory is something of an anachronism. As automakers strayed from their vertically integrated origins, some concluded they were little more than aggregators or integrators of already-constituted units, no longer performing much of the actual labor that vested their vehicles with value. So complex were these arrangements that many automakers developed their own in-house logistics firms, supplementing or replacing third-party logistics, to better manage the shipment and transfer of modules and subcomponents under the unforgiving timetable of just-in-time production.

Today's auto factories are technically termed final assembly plants because the work consists of fitting together a series of already-finished parts rather than creating an automobile out of whole cloth (steel). Like most contemporary auto plants, VW's Chattanooga operation relies on a combination of offshoring and outsourcing for the manufacture of components. Parts must be sourced from third-party subcontractors, most of which have no formal relationship with the purchasing firm. Sophisticated parts—such as engines—often have tertiary and quaternary suppliers who themselves enter into agreements with the engine manufacturing firm. This arrangement introduces layers of complexity and logistical challenges on a new order of magnitude, giving rise to the growing industry known as supply chain management. Meanwhile, worker power is increasingly fragmented and dispersed.

Insourcing (Vertical Reintegration)

Yet the fact that VW outsources much of its productive capacity does not mean suppliers are free from its oversight. VW presides over a vast network of contractors, some of which are essentially VW owned in all but name, serving as the company's exclusive favored suppliers and, in some cases, barred from bidding on competitor contracts. The fact that these workers are not counted among VW's U.S. workforce, and their plants are not consistently seen as a crucial component of any organizing initiatives, is a mere accident of corporate accounting policies.

Thus, even though its suppliers may not be in-house, VW has tried to keep them nearby and under its control. Well before it invested in Tennessee, VW already had a reputation for keeping its suppliers on a tight leash—a policy the firm largely maintained even as it expanded its global operations. Yet despite the ready availability of parts makers in Tennessee as of 2011, protocol barred VW from contracting with certain firms that had pre-existing arrangements with the state's Nissan and GM plants, requiring it to quickly recruit a new supplier base. As it opened its Chattanooga plant, VW constructed two massive buildings across the street, which effectively constituted an in-house supplier park that eventually employed 1,100 workers.[36] Local firms produced key components, including bucket seats and drivetrains. In a press release, VW described its supplier park using rhetoric that might have been pulled from a logistics white paper:

> Every mile of distance a supplier needs to haul their parts can add hundreds of thousands of dollars in annual costs. Just-in-time production systems like the one used at VW plants worldwide help keep on-hand inventories low so that workers get the parts just as needed. Suppliers can work more efficiently by being as physically close to the plant as possible.[37]

The purpose-built supplier park—clustering taken to its logical extreme— also has some support in the literature.[38] Tennessee has the highest concentration of automobile jobs per capita of any state, with most of these jobs centered in the dozen counties that surround its three assembly plants—all of which are within one hundred miles of one another. State policy encouraging clustering has resulted in an exceptionally high rate of in-state supply purchases by manufacturers.[39] As of 2011, the state had sixty-eight suppliers, more than any other southern state, and employed seventeen thousand workers (also more than any other southern state, surpassing its nearest competitor by 40 percent).[40] To be fair, some of these suppliers are exclusive to Nissan and GM plants, which predated VW by twenty years, but many others directly supply VW or benefit from regional economies of scale (the full details of VW's supply chain are proprietary).

Economic impact models also indicate the importance of supply chains. Economic development boards salivate over auto plants partly because of their demonstrable secondary and tertiary effects on the local employment base. One estimate suggested that 3.49 additional jobs are created via multiplier effects for every auto industry job.[41] A 2013 study by the University of Tennessee concluded that when VW Chattanooga is positioned alongside its regional suppliers and multiplier effects, the plant accounts for $483.9 million in income and some 9,985 jobs—above and beyond the 2,415 people employed at the factory at the time.[42] Both the city's Office of Economic Development and the Chamber of Commerce expressed particular fealty to VW because of its multiplier effects. While VW ranks third among the city's largest employers, its supply chain effects easily send it to the top of the rankings.[43] Including these jobs, as much as 5 percent of the city's population is employed directly or indirectly by VW.

Beyond the supplier park, VW had courted a resplendent supplier base with cooperation from local, county, and state officials. At the plant's inception, VW counted fourteen suppliers with two hundred or more employees within a one-hundred-kilometer radius. The efforts have proved effective. VW's plant in Chattanooga receives 87 percent of its transport material from the United States and Canada as of 2016, and sources 60 percent of its total supplier purchases in the state. However, there is significantly less clustering further down the value chain. Second-order and third-order suppliers often had much longer supply lines. For example, Chattanooga Seating Systems receives 87 percent of its material from Mexico.[44] The dispersal of the supply chain at its lower levels is a direct result of competitive pressures from above.[45]

Stratification of the Automotive Workforce

The previous sections have shown how the reorganization of production weakened worker power. But solidarity is even more difficult to sustain when the sector is internally divided along lines of employment status.

Tiers

The UAW initially approached VW workers in Chattanooga from a position of weakness. At its unionized assembly plants, workers were segmented into endless divisions and subdivisions based on their relative skill, hire date, and employment status. Tiers and quasi tiers varied enormously in pay rates, benefit eligibility, and job security, further undermining industrial solidarity by creating artificial hierarchies and fostering competitive pressure over scarce resources. Moreover, suppliers, subsidiaries, and distribution centers functioned as de facto "tiers," with substandard wages despite comparable work.[46] Until the 2023 contract, when the UAW vowed to eliminate tiers, the union had been reduced to negotiating within the tier system, conceding management's right to pit subgroups of workers against one another for a share of an ever-shrinking pie. Indeed, far from eliminating tiers, as negotiators initially promised, pre-2023 deals deepened divisions within the UAW's ranks, splitting workers into microdivisions up and down the value chain.

Key to the UAW's historic success was the principle of *industrial unionism*—the idea that comprehensive union contracts should cover all workers, independent of skill, seniority, or job classification. By the time the UAW reached its peak membership in 1979, there were already signs that the era of shared prosperity was coming to a close. The oil crisis of the 1970s gave perhaps the first outward indication that the postwar labor peace was on shaky footing. Rising prices for gasoline led consumers to favor smaller, more efficient Japanese models instead of overbuilt American models, resulting in a significant loss of market share for the Big Three, slimmer profit margins, and cycles of layoffs in legacy auto plants.[47] While the UAW has retained its hold on all of the Big Three's assembly plants (including those in the South), both its raw numbers and its industry-wide density figures have been reduced by a series of factors, including the offshoring of plants, the spinning off of parts suppliers (often to nonunion firms), and, most relevant to our discussion, the rapid rise of foreign-owned auto plants on U.S. soil. Significantly, by 1979, the UAW had broken its decades-long tradition of pattern bargaining, trading some $200 million in concessions at Chrysler in a pact to keep the company afloat, and setting a dangerous precedent that would haunt it for years to come.[48]

The 2007 bargaining round marked the final collapse of whatever remained of the social contract and, with it, the principle of industrial unionism. That year, the UAW agreed to shift the cost of health care from employers to a semi-independent, UAW-administered health plan, which has since suffered from chronic underfunding. This contract also included a provision to hire new workers at substandard pay, requiring these entry-level workers to be capped at 20 percent of the overall workforce and limited to "noncore" areas of the plant. This restructured pay scale—known as *two-tier*—was met with derision by activists both because it reduced pay and because it created an artificial division between workers, undermining in-plant solidarity and potentially reducing union power.[49]

If two-tier was originally intended as a simple carve-out designed to reduce costs while protecting the UAW's core constituency, the economic crisis in 2008–2009 signaled that two-tier would have staying power. Though UAW president Ron Gettelfinger resisted restructuring in the waning days of the George W. Bush administration, the UAW eventually agreed to reopen its contracts under pressure from the incoming Barack Obama administration to save GM and Chrysler from near-certain collapse. As Gettelfinger discovered, once the Pandora's box of tiered contracts had been opened, it proved almost impossible to close and was even expanded in the government's bailout package. Negotiations to revive the bankrupt GM resulted in further setbacks for second-tier workers, who faced no path to full pay, a limited pension, and a six-year phase-in before receiving raises and bonuses. Moreover, the 20 percent cap was lifted (though some restrictions remained), pushing the percentage of second-tier employees as high as 40 percent in some plants.[50]

Temps

While VW had no official tiers, its system of temps served an analogous function. Nearly all new production line hires started as temporary employees, then advanced to regular employment after a six-month probationary period.

Temps worked alongside permanent employees and performed the same tasks as direct hires for 25 percent less pay. Though VW does not publicly release data on the internal composition of its workforce, estimates placed temps at 30–40 percent of the total workforce during the plant's early years.[51] (These numbers fluctuated wildly depending on production cycles and could easily be adjusted to match demand.) Though VW incentivized temp work with the promise of full-time conversion after a minimum of eighteen months of temp service, burnout was common, and conversion to full-time work was purely at the company's discretion. Further, though neither permanent workers nor temps enjoyed any meaningful job security, temps were subjected to even harsher working conditions than direct hires to "weed out weaker team members."[52] For all these reasons, most temps never received full-time status, languishing in interstitial "permatemp" status or quitting before they became eligible.

It may be tempting to view VW's temporary employees as a class apart from direct hires, necessarily excluded from representation by virtue of their tenuous employment status. However, several factors complicate this picture. Inside the plant, temps were virtually indistinguishable from their directly employed counterparts, who shared the same spaces and performed the same tasks (except for jobs requiring special training or technical expertise). Moreover, after the initial hiring wave, VW channeled all workers through temporary employment purgatory, so most temps saw themselves as permanent workers-in-waiting, while nearly all full employees had previously graduated from the temp status. Thus, the strict distinction between temporary and permanent employees failed to capture the authentic experience of the division of labor on the shop floor.

Yet temps provided management with some clear advantages. While temps received some benefits, their package was not comparable to that of full employees: most notably, they were not permitted to lease a company-subsidized car, and their health insurance was a cheap knockoff of the VW plan. When viewed through the lens of total compensation, the cost savings between permanent workers and temps was significant, even after the subcontractor collected additional fees for its hiring and screening services. In this sense, temps provided VW with a second layer of preemployment screening: those who could not meet standards could be let go at no real cost to the firm.

Flexibility is an important part of the lean production playbook because it creates competitive pressure between temps and mainline employees, using each as a check against the other's power. Lest workers grow too confident, the temps serve as a reminder of the ever-present possibility of layoffs. Temps, for their part, look toward mainline employees with resentment tinged with jealousy—the ultimate motivator. If the contractual relation between boss and employee was founded on a codependence that often veered toward paternalism, the rise of temp work has rendered workers an undifferentiated, disposable mass, depersonalized and interchangeable.

Conclusion

This chapter has situated the Chattanooga Assembly Plant in the context of the restructuring of the American automobile industry and the decline of the UAW. By 2014, the UAW confronted an industry that had abandoned its historical home, embraced the mythos of the "New South," partitioned itself out among several small-holding suppliers, and come to rely increasingly on temporary work. In short, labor was divided and compartmentalized by region, firm, and employment status, undermining solidarity and contributing to plummeting unionization rates (even if just-in-time production and strategic recoupling pointed toward new opportunities). But as debilitating as these macroeconomic trends were, they cannot fully account for the 2014 election outcome in Chattanooga. To understand these events, we must examine how VW management strategically leveraged industry-level trends and collaborated with regional and national forces to defeat the UAW.

2

SOLIDARITY DENIED

How the UAW Was Defeated in 2014

This chapter introduces our case study and provides a roughly chronological narrative description of events in the period from the first public rumblings of the UAW's organizing drive in 2012 through the February 2014 plant-wide election.

The UAW's 2014 organizing strategy hinged on expectations that VW might be a soft target that would concede to unionization without a real fight. Early on, the UAW even hoped the company might voluntarily recognize the union on its own accord. However, those expectations quickly proved overly optimistic, as VW's position on unionization was rife with contradictions. In general, German executives favored carrots (union substitution), while American managers and supervisors preferred sticks (union suppression). Moreover, management was a moving target, alternating between a conciliatory and an adversarial approach to the UAW as the campaign developed. Therefore, instead of scoring the anticipated easy win, the union locked horns and dug in for a multiyear standoff, facing off against both the company and a coalition of oppositional forces arrayed against it.

Despite VW's positive labor relations in Germany, the UAW faced an uphill battle in Chattanooga. Table I.5 in the Introduction of this book shows that while only some of these challenges would recur in 2019 and 2024, each held valuable lessons the union would draw upon over the decade. First, VW held early appeal as a job creator and harbinger of further investment in an industry-starved region. Leaning heavily on its image of technological sophistication, the company seduced workers and generated initial goodwill in the community through *the honeymoon effect*. Second, the company resorted to *promises and improvements*, including minor changes to working conditions, in an attempt to ward off the union threat. Third, shop-floor supervisors engaged in overt *union suppression* activities, including threatening, intimidating, and disciplining suspected union supporters. Fourth, management employed *delay and stalling* tactics, creating uncertainty and frustration among workers. Fifth, *political interference* took the form of a concerted campaign to demonize the UAW, with an in-plant "astroturf" component. Sixth, the *threat of disinvestment* or relocation played to workers' fears of job loss. Seventh, VW's German executives favored a *union substitution* strategy, which attempted to provide benefits and worker representation through alternative means, thereby reducing the perceived need for a formal union. Eighth, the instability and volatility of VW's corporate culture, including a perceived disconnect between headquarters and overseas production, contributed to what I term *deflection and distancing*.

Parenthetically, I note that this analysis treats *political interference* and the *threat of disinvestment* separately. Dire warnings and threats from "antis" were a constant drumbeat across all three elections, and their message was remarkably consistent each time: a unionized plant would be prohibitively expensive, forcing the company to reduce its capital outlays through layoffs or divestment from Chattanooga. Because of the comparative nature of this study, it emphasizes changes that affected the organizing environment and influenced the UAW's prospects across the three elections. The disinvestment threat was persuasive in 2014, moderately convincing in 2019, and out of touch to the point of irrelevance in 2024. Thus, political interference remains constant, even as the threat of disinvestment dwindles.

The Honeymoon Effect:
Novelty and Boosterism

VW generally cooperates with unions abroad, and there were reasons to believe VW's culture might arrive in the United States more or less intact. Its early conduct signaled an intent to preserve its German production model rather than changing its colors to match the scenery. Specifically, in a move almost unheard of in the American context, VW positioned itself early on as a benevolent corporation, eager to accommodate employee representation, at least in some form, with the intent of eventually organizing a works council in the model of its German plants.

VW has long benefited from an image of technological sophistication stereotypically associated with German companies. In the 1990s, its advertising agency famously coined the fake German word *Fahrvergnügen*, connoting an exotic, inscrutable expertise (much to the amusement of its home-country workforce). More recently, it adopted the tagline "German engineering" to remind consumers of the vehicle's provenance, even as the firm became increasingly globalized. The mythology of the VW brand proved seductive. These ideas featured prominently in the company's internal culture and aligned nicely with Chattanooga's efforts to rebrand itself as a forward-looking city.

VW came to Chattanooga pledging to be a different kind of company with high-minded social goals, strong protections for workers, and a deep commitment to the community it now calls home. Not everyone was equally entranced by VW's glittering promise, but the company's public relations blitz was undeniably seductive, luring prospective employees and percolating down to the shop floor, at least during the first years of the plant's operations. Well before the first chassis rolled down the production line, VW was already receiving ample press coverage as it began to mold the region's unformed masses into future citizens of industry. The local newspaper ran regular features on VW history and company culture alongside updates on plant construction, all with the goal of introducing Chattanoogans to their new corporate neighbor. The tone of these pieces was ecstatic, less balanced journalism than breathless boosterism.[1] They conveniently ignored or glossed over dark chapters in VW's history, such as its perennial difficulties with Mexican trade unions. As with the open-armed reception that greeted foreign automakers elsewhere,[2] these media accounts served as a kind of *anticipatory socialization*, preparing potential employees to take on new roles and raising their expectations before recruitment even began.

Several employees described VW's relatively generous pay and benefits as a principal factor working against support for the UAW. The standout component of this package enabled employees to purchase a vehicle directly from the plant at a drastically discounted rate, thereby bypassing third-party dealers. Given that personal automobiles are major status symbols in the South, and given that the Chattanooga area is deeply imbued with car culture, this perk had symbolic value beyond its monetary significance. The personal automobile is deeply intertwined with American values of freedom and manifest destiny, which have particular currency in the South, where mass transit and sidewalks barely exist, encouraging the individualistic fantasy of the open road.[3] The employee purchase discount extended company loyalty to workers' immediate families and deepened investment in the brand. Under this policy, VW workers, often the principal breadwinners in their households, could provide their loved ones with that quintessential marker of cultural inclusion: vehicular mobility.

Even those who rejected VW's efforts to cultivate loyalty and saw through the company's class-blind camaraderie tended to recognize that their city's fate was bound up with VW's prosperity. As with other transplants, VW invested heavily in an effort to increase the sense of local "embeddedness."[4] A locally made prototype vehicle greeted visitors just beyond the security gates at the regional airport. The spirit of boosterism was almost inescapable in Chattanooga, from the ubiquitous logos to the company-branded professional soccer team; to be anti-VW was to be anti-Chattanooga. Therefore, workers were happy to criticize specific policies but largely refrained from attacking the company itself, which they saw as a net benefit to their city.

As a result, the response to VW's initial hiring call was enthusiastic. As luck would have it, the plant benefited from downsizing at the Nissan plant outside of Nashville and from the idling of a Ford and GM plant in the Atlanta area—none of which were known or predicted at the time of the Chattanooga plant's conception.[5] Both developments proved fortuitous, providing Chattanooga with a ready supply of experienced autoworkers within reasonable commuting distance. The company reported receiving over sixty-five thousand applicants for two thousand production jobs, allowing it to be highly selective.[6] Plant president Frank Fischer also gleefully noted that the city and state had placed no "preconditions" on the hiring process, so the company could safely reject at-risk populations, "like the long-term unemployed."[7]

Chattanooga's rapid transformation into an auto production center begs comparison to World War II, when the rapid expansion of industrial production to meet wartime needs necessitated the recruitment of female, Black, and rural workers, many of whom were new to both factories and unions. The absence of a union tradition among these new workers resulted in an interesting paradox: a high level of shop-floor militancy and conflict but a complete lack of interest in constructive union-building activity. In practice, this played out in a wave of wildcat strikes that extracted some short-term concessions from employers but did little to further the long-term prospects of the UAW. The insurgent masses that flooded the factory gates did not lack class consciousness, but in their brazen militancy, they saw no need for the methodical deliberation involved in organization building.[8] Admittedly, the comparison is imperfect. The first cohort of workers at VW never came close to considering a wildcat strike, but their activism sometimes took on a defiant, antiauthoritarian character.

Anecdotal evidence suggests, however, that VW shied away from hiring many former Ford or GM workers from recently shuttered plants in the Atlanta area, who bore the stain of their unionized past. Experienced workers were desirable, but experienced unionists were not. It should be noted that a factory job paying relatively low wages located two hours away may have been less than enticing to former Ford workers from Atlanta, accustomed as they were to union wages that ran 30 percent higher, depending on experience and classification. Still, ex-Ford workers were conspicuous in their absence.

VW seemed willing to hire workers from nonunion transplants in the Sunbelt, but the UAW's unionized southern plants were pointedly shunned.

VW's relatively generous pay scale also contributed to the company's early appeal. If Chattanooga's hourly rate paled in comparison to VW's international standards, it looked good in the context of a provincial southern city with a devastated manufacturing base. Chattanooga had once been a southern industrial hub, but two of the last large factories in the area—Wheland Automotive and U.S. Pipe (both unionized)—had closed in the decade before the plant's induction. Since the implementation of the North American Free Trade Agreement (NAFTA), local manufacturing jobs had declined by 268,000, including 60,000 unionized jobs.[9] As is typical of a greenfield (newly constructed) auto plant, VW paid somewhat more than the market might dictate. However, with the meager wages in the Chattanooga area, this bar was set artificially low.[10]

Many workers claimed that pay was a net positive, especially for younger employees without financial independence from their parents. Blue-collar workers at VW earned considerably more than comparable *industrial* laborers in the region, even as Chattanooga's mean hourly wage was about 15 percent below the national average.[11] This relative advantage helps explain why the prospect of a VW career generated so much interest and enthusiasm, despite the regional pay differential compared to Detroit. The pay scale was appealing to non-college-educated workers at the margins of the economy, for whom upward mobility was unlikely. Other scholars have observed that above-market wages at transplants block proletarian self-identification, causing workers to imagine themselves as "chosen" elites instead.[12] In this view, the sheer desperation of southern workers blinds them to the conditions of their exploitation.

Promises and Improvements

The company also warded off the union threat by seducing supporters with its pro-worker reputation and making minor improvements to working conditions, most of which would not outlast the union drive. From the company's perspective, these tactics were enormously effective, lending credence to the "catch more flies with honey" adage. While anti-unionism is dogma to most foreign companies operating in the United States, VW was uncharacteristically cooperative. During the 2014 election cycle, senior management refrained from overt, heavy-handed anti-union tactics. For example, the firm did not employ labor consultants (specialized advisers who train managers in union-busting techniques), as is standard practice in contested elections in the American private sector, nor did it hold mandatory captive audience meetings. Breaking again with the dominant pattern, it allowed union representatives to campaign freely in its factory, met regularly with union leaders, and maintained an official position of neutrality with respect to the election. The two top management representatives at the plant even wrote a letter to workers that spoke glowingly of the UAW and described the proposed path from unionization to the works council as an "innovative model."[13]

Early in 2012, nearly two years before the election, the company suggested it would recognize the union voluntarily through the card-check procedure. Card check is a common strategy that may benefit both unions and firms, but it depends on the company's cooperation under U.S. law.[14] In the last decades, the service-sector "organizing unions" of the former Change to Win Federation have enjoyed success through reliance on card check, but former CIO-affiliated industrial unions have been slower to appreciate its advantages. Within the UAW, the card-check strategy was less common, though not unprecedented; the UAW had successfully organized a Freightliner plant via this method just months earlier.[15] Among many other benefits, card check is seen as a means of bypassing the opposition, which can derail a conventional union election.[16] In an offhand comment that would prove unexpectedly prescient, UAW Region 8 director Gary Casteel justified the card-check strategy by saying, "We know if we go for a traditional election where the outside organizations could campaign against us, we would probably lose."[17]

In March 2013, Horst Neumann, VW's leading human resources (HR) executive, made headlines, acknowledging that the company had initiated talks with the UAW about the possibility of unionization as a necessary precursor to a works council, implying that the company would allow a certification petition. Since traditional union elections require no company foreknowledge and typically operate under cover of secrecy, union-company coordination at this (early) stage of the process implied that card check was a likely outcome.[18] Given that VW had a fifty-year record of negotiating with its works council and bargaining with works council–affiliated unions in good faith, there was reason to believe that past precedent would set the tone for VW's engagement with the UAW. As it happened, the company's early pro-worker posturing would eventually prove a ruse, but not before it had seduced the union and averted a serious organizing campaign.

Union Suppression

VW surprised many with its official position of neutrality in 2014. However, the company's official position obscured the reality of recurring, pervasive anti-union behavior by supervisors. Despite the apparent wishes of their German superiors, shop-floor managers often intimidated and harassed workers or sought to undermine organizing efforts. It is unclear whether this behavior was actively encouraged, tacitly permitted, grudgingly tolerated, or simply unbeknownst to VW higher-ups. However, a preponderance of evidence indicates that supervisors resorted to vigilante-style union suppression tactics, often without censure from superiors. They systematically employed threats, intimidation, punishment, and surveillance—behavior that meets the standard for unlawful anti-union conduct. Even when the UAW flagged its actions, senior management maintained a policy of malign neglect and took no steps to rein in its supervisors.

These supervisors, operating in their official capacity but apparently in defiance of the company's directives, resisted the union using legal and illegal means. They sometimes used job assignments to reward and punish workers for their pro- or anti-union attitudes. Supervisors could (selectively) inform their superiors of employee misbehavior, often resulting in corrective action up to and including dismissal. These activities were well documented and fairly consistent across multiple accounts.

Still, there were limits to this conduct: On at least a few occasions, supervisors were reprimanded for their anti-union conduct. But more typically, supervisors engaged in anti-union behavior with relative impunity. In the eyes of one worker, supervisors steeped in American-style anti-unionism were blindsided by a company culture that had traditionally been union tolerant. If senior management was relatively receptive to unionism, the mostly American supervisors took the opposite tack. In their view, unionization was a threat to the arbitrary and nepotistic system of informal control they had come to enjoy.

Such harassment was effective from the company's standpoint in that it had a chilling effect on organizing. Since nearly all workers were novice organizers, and because the union's efforts to inoculate workers against potential reprisal were minimal, they were ill equipped to confront supervisors when their rights were violated.

Even when union supporters did not face immediate repercussions, a culture of surveillance discouraged them from open advocacy. While the UAW reported incidents of harassment and mistreatment to higher-ups in Germany, these managers were not disciplined or removed, and the union did not file unfair labor practice complaints (ULPs) at this stage, fearing it would antagonize the company.[19] The UAW's determination to maintain good relations with the company led it to downplay such concerns rather than demand that the company take aggressive action against offenders.

Delay and Ambivalence

Though overt union suppression by supervisors drew the most attention, senior management deterred unionization through "passive" or "tacit" means, including delays, inconsistencies, and general ambivalence.

The union signaled it had achieved the requisite majority support in the summer of 2013, but there were immediate signs that voluntary recognition would not occur quickly, if at all. The backsliding started when HR director Frank Patta called an "all-team meeting" to beat back anticipation and stave off uncertainty about achieving union representation. Noting that the process would be lengthy, Patta warned that "legal issues" would prevent any swift resolution while pointedly ignoring that the chief obstacle to representation was the company's refusal to immediately certify the petitions.[20] The company had never provided the union with a list of eligible workers, nor had it disclosed the size of the proposed bargaining unit. Union supporters continued to hold out hope that VW might voluntarily recognize their union, but fate (or the company's fiduciary obligations) would dictate otherwise, and as the weeks turned into months, the already tense climate only grew more acrimonious.[21]

Weeks later, the prospect of card check loomed large during high-level talks between UAW and VW executives, including UAW regional director Casteel, head of HR Horst Neumann, and then-head of the Global Works Council (GWC) Bernd Osterloh. (Employees received no advance notice of this event and were only made aware of it retroactively through the German press.)[22] The trilateral summit had all the trappings of a moment that might pave the way for union recognition. The company never disputed the extent of the union's support and, at this juncture, could have voluntarily recognized the union, preemptively avoiding what would later become known as the Battle of Chattanooga. There is no record of what transpired in this meeting, but by the next day, VW's head of U.S. operations declared the union would only be approved through a "formal vote."[23]

Waiting for the card check that never came cost the union crucial months that would have been better spent campaigning. But this was not the first time the UAW had banked on card check, only to watch the prospect implode.[24] Neglecting to learn from its experience, organizers in Chattanooga felt deeply betrayed in the fall of 2013 when the company rejected card check after previously indicating it would accept that process.

After the company reversed itself on card check, it agreed to issue a neutrality statement, meaning it would not interfere with the election. Union supporters viewed a commitment to neutrality from VW as a necessary prerequisite for a fair union representation election. However, the neutrality statement was endlessly postponed, leading some workers to conclude that the company was less than sincere in its commitment. As late as January 24, 2014, two weeks before the vote would eventually be scheduled, VW was still not forthcoming with a promise to remain neutral. When it belatedly emerged, the company's neutrality statement ("Agreement for a Representation Election") had little relation to management's actual behavior. Once again, this resonated with the UAW's previous experience with other transplants. In 1982, the UAW secured a neutrality pledge from Honda, only to watch as it was later abandoned under pressure from community groups. The company continued campaigning against the UAW, forcing the union to withdraw its petition. Similarly, the UAW's efforts to leverage Mercedes's position as a subsidiary of unionized DaimlerChrysler to its advantage during organizing drives in 1999 and 2000 were defeated by a clandestine anti-union campaign by supervisors.[25]

Political Interference

Time is an organizing drive's worst enemy. As a result of the delay tactics discussed in the previous section, the ramp-up to the vote lasted nearly eighteen months, providing the opposition with ample opportunity to gather its resources, coalesce, and broadcast its message to a receptive audience. As the election process dragged on, a coterie of political opponents had already begun to organize behind the scenes. The endless delay gave them a window of opportunity to mount an opposition campaign that quickly overshadowed the union's meager efforts to that point. The union was surprised and unprepared for the emergence of this campaign, but it ought to have anticipated opposition, given its past experiences.

In a typical union drive, the employer funds, coordinates, and directs the anti-union opposition. That did not occur in this case: to maintain its neutral stance, VW did not retain anti-union consultants and did not intervene directly in the opposition campaign. Senior management remained above the fray even as the climate grew more heated. Instead, the public-facing opposition campaign was left to "outside agitators" backed by an array of national groups—many with deep pockets—mounted an anti-union effort as formidable (if not more so) as any the company might have devised on its own. By late 2013, Chattanooga had become a cause célèbre for the anti-union set. One anti-union consultant noted, "Everyone is looking at this fight. Everybody is looking to play their part and get compensated for playing their part."[26]

The opposition was spearheaded by the efforts of Matt Patterson, a senior fellow at the Competitive Enterprise Institute (CEI), who took on the UAW as a pet project. The CEI is a nonprofit libertarian think tank that has emerged as one of the most influential defenders of unbridled capitalism, known especially for its anti-union stance. Patterson's efforts included placing anti-union editorials in local media, renting a billboard with the message "Auto Unions ATE Detroit. Next Meal Chattanooga?" and rallying Chattanooga's far-right community groups. As it turned out, his messaging was borrowed almost verbatim from union opponents at Mercedes, whose billboard read, "Don't let Alabama Turn into the Next Detroit."[27]

The origins of this anti-union effort were murky at the time and would only be fully exposed after the election. Patterson maneuvered to conceal his contributions behind locally owned shell firms, giving the opposition campaign a measure of "grassroots" credibility. Though local business interests—possibly even including VW itself—contributed money toward his anti-union activities, a series of leaked documents revealed deeper influence.[28] By August 2013, Patterson had attracted the backing and funding of Grover Norquist of Americans for Tax Reform, who granted Patterson's efforts "special project" status and anointed him executive director of the newly formed Center for Worker Freedom (CWF). With Norquist's backing, some two thousand autoworkers living in relative obscurity in the country's 137th-largest metropolitan statistical area were thrust reluctantly onto the national stage.

Norquist was a master of media manipulation, and his messaging was tightly controlled and fairly consistent: unionization would kill jobs and reverse Chattanooga's economic ascent, with ripple effects well beyond the auto industry. CWF-backed pundits predicted that unionization would lead to plant closures,[29] cost peripheral auto jobs,[30] and "destroy" other auto plants across the South.[31] One commentator even conjured the image of the invading Union army.[32] These efforts were complemented by the involvement, starting in September 2013, of the National Right to Work Legal Defense Foundation, which coordinated and financed the legal component of the anti-union campaign, sometimes in collaboration with sympathetic local attorneys. The local newspaper featured a series of editorials comparing the union to a parasite, a cancer, an invading army, and other villains, real and imagined. A sampling:

> A union in a company acts very much like a virus in a body; even if the virus itself is not fatal, it can leave its host body weakened and drained of the resources it needs to survive.[33]
>
> I'm reminded of an insect that reproduces by identifying a host and injecting its egg inside. The Ichneumon wasp larva feeds parasitically until its host is killed.[34]

The UAW did not have a real media strategy until much later, so these claims went unanswered.

Politicians complemented the dark money now streaming into Chattanooga with an anti-union offensive of their own. Typifying their hypocrisy, forces led by Chattanooga's right-wing ruling class browbeat the UAW for its insider dealing, but their representatives were equally guilty of the same. Even as UAW leaders were jet-setting to Wolfsburg for high-level meetings, leading Republicans were secretly doing the same. If the UAW had VW's full attention through the summer of 2013, Tennessee politicians had begun to assert their provenance by that fall, holding closed-door meetings with key VW executives in the governor's mansion.[35] Though the full extent of VW's aggressive rent seeking and the state's largesse was unknown until after the vote, it has since been revealed that Tennessee politicians dangled $300 million in additional subsidies to VW with the implicit proviso that the offer could be rescinded if the plant went union. This offer was first extended on August 23, 2013, but was withdrawn on January 31, 2014, as the union vote was still pending.[36] The payment was finally made in March 2014, once the union situation had been resolved, and became publicly known only in July 2014. The state was leading the company by the purse strings. Though this revelation provided further evidence of unlawful vote manipulation (and fodder for additional legal challenges), it was already a done deal once the money trail was exposed.[37] While Haslam's representatives tried to cover their tracks, the timing of these events proved that Haslam's team had tried to manipulate the vote's outcome through a bait-and-switch operation. Though it is not unusual for states to incentivize plant expansion through subsidies, the strategic use of subsidies to influence a union election is a brazen move, even by the standards of the anti-union playbook.

The in-plant opposition first revealed itself through a multipronged operation that included shop-floor, legal, and publicly oriented strategies; at times, it resembled a Ray Rogers–inspired corporate campaign.[38] Anti-union workers coalesced under the banner "Southern Momentum," a name evoking down-home roots while also suggesting Tennessee's ongoing resurgence. The UAW, in contrast, was portrayed as a Rust Belt relic still unable to escape its late twentieth-century doldrums. First came the debut of the anti-union website no2uaw.org.[39] Days later, several disgruntled employees submitted a complaint alleging they had been compelled to sign authorization cards under false pretenses and demanding that their signatures be revoked. Later, the same group submitted a counterpetition asking the National Labor Relations Board (NLRB) to throw out the UAW's cards, eventually persuading 30 percent of the workers to demand that their signatures be voided.[40] Simultaneously, a lawsuit alleged that the UAW's cooperation ("collusion") with VW constituted a labor law violation.[41]

In sum, during the eleven-month lead-up to the 2014 election, the opposition had the opportunity to coalesce and mount a formidable challenge. Though never large in number, the in-plant opposition took advantage of the power vacuum and presented itself as the authentic mouthpiece of the rank and file. Without a proportionate response from the absentee UAW, these claims went unchallenged.

Disinvestment

Research has shown that management threatens to shutter plants in 70 percent of manufacturing facilities where a union drive is underway.[42] Chattanooga was no exception. Concerns over the plant's viability were present from the beginning as it struggled for years to ramp up production and meet designated targets. During its first decade of operation, the plant achieved its projected output of 150,000 vehicles only once, in 2012, during its first full year in production.[43] Though the workforce's size generally increased in the following years, periodic layoffs punctuated the upward trend.

In March 2013, roughly a year before 2014, five hundred temporary workers employed by the staffing agency Aerotek were unexpectedly laid off. Auto plants commonly cycle through workers as seasonal demand fluctuates. In fact, the arrival or departure of temps was among the best indicators of the plant's overall performance. Perversely and without basis, the union took the fall for these layoffs, as rumors circulated that layoffs among Aerotek workers were partly fueled by the union's presence. While these rumors were unsubstantiated and the cuts were likely attributable to the quality issues typical of a newly constructed plant, they had a dampening effect on organizing.[44] Layoffs of Aerotek workers, therefore, added to the sense of uneasiness among full-time employees. Workers had similar experiences at other transplants, such as Mercedes, where temps "increased the insecurity of the permanent workforce."[45] Although Aerotek workers were excluded from the bargaining unit, in practice, their daily duties were not much different from those of direct hires. But workers saw Aerotek layoffs as an implicit threat to their own jobs. These layoffs were accompanied by unfounded rumors about a plan to outsource logistics jobs.[46] The appeal of temp work lies partly in the ability of firms to dial up or ramp down staffing levels as production needs fluctuate, so these layoffs were not especially revealing, but they contributed to a generalized climate of uncertainty.

Though the plant's design features drew accolades, the company's sales forecasts in 2014 were less rosy. Apart from its one-off success with the counterculture icon Beetle some forty years prior, VW had never achieved much market share in the United States, and it suffered from a perception of low quality and poor reliability.[47] VW's struggles in the American market were somewhat predictable, as it boasted neither the fuel economy and low price point that made Japanese models attractive nor the romantic pioneer spirit and patriotic devotion that buoyed the Big Three even as foreign rivals outpaced them. Moreover, VW had been slow to develop the SUVs, crossovers, and all-terrain vehicles that were increasingly popular in the United States.[48] Depressed fuel prices and consumer preferences for oversize vehicles made technically impressive but physically underwhelming German cars a poor fit for the

American market. But it would be six years before Chattanooga was assigned to produce a large vehicle; its first order was the Passat midrange sedan, which was never a strong performer in the U.S. market and sometimes drew derision for its diminutive profile. VW's decision to locate its newest assembly plant in the United States was widely seen as an effort to boost its popularity among American consumers. Here, it stole a page from Toyota, whose Kentucky assembly plant—backed by an aggressive "Made in America" marketing campaign—helped destigmatize the firm among xenophobic buyers.[49]

Apart from fears that VW might reduce its footprint in Tennessee, unionization was seen as a threat to VW's ongoing investment in the region. The VW plant had been constructed with room to expand, and it had long been anticipated that Chattanooga might eventually be awarded a second, more lucrative model. While such an expansion was never affixed to a timeline, VW put the decision to expand on hold amid the UAW drive, pending the vote results. Stephan Wolf, deputy chair of VW's Wolfsburg plant works council, said, "We will only agree to an extension of the site or any other model contract when it is clear how to proceed with the employees representatives in the U.S."[50] Wolf spoke as a works council representative, but since the council has seats on VW's executive board, his statement appeared to carry the imprimatur of VW proper. This statement only indicated that VW intended to delay its expansion announcement until after the contentious vote, without explicitly suggesting that the two events depended on each other. But for anti-union opponents reading the tea leaves, this indicated that VW's planned expansion hung on the vote's outcome.

Haslam and Corker moved quickly to tie the plant expansion to the unionization vote—initially through implication and conjecture and later through direct and explicit threats. As early as spring 2013, Haslam was already equating unions with job loss: "I would hate for anything to happen that would hurt the plant's productivity or deter investment in Chattanooga."[51] But the night before the vote, Corker dropped a bombshell, flying back from Washington to tell workers:

> I've had conversations today and, based on those, am assured that *should* the workers vote against the UAW, VW *will* announce in the coming weeks that it will manufacture its new midsize SUV here in Chattanooga.[52]

Corker's rhetoric was familiar, even tired, but his *if-then* causal logic was in a different register than past threats. Also, for the first time, it incentivized the vote with a carrot (*vote no and expand the plant*), not a stick (*vote yes and lose your job*). Corker's shocking revelation dominated the news cycle in the crucial window before the vote. VW officials were measured in their response, refusing to be baited by Corker while being a bit too eager to note the importance of reducing production costs in preparation for the new vehicle.

Other politicians took the disinvestment threat further, stating that fallout resulting from a yes vote would extend beyond the plant to the regional supplier base. They worried a successful campaign by VW's production workers would tarnish the city's pro-business image. Mark Sweeney, senior principal of the site selection firm McCallum Sweeney Consulting, spoke for many, suggesting that the viability of the entire Southeast megasite could be jeopardized should the UAW win.[53] Concurrently, state senator Bo Watson took to the airwaves to predict that a yes vote would result in the denial of future state subsidies to VW, an implied blow to the company's expansion plans.[54]

While the opposition's fearmongering is rightfully stressed in most accounts of the vote outcome, few have noticed that pro-union forces sometimes engaged in parallel games of high-stakes brinksmanship.[55] The works council fed the same type of thinking from the other side of the aisle, insinuating without basis that a "no" vote would cause VW to second-guess its expansion plans. Immediately after the vote result, GWC chair Osterloh hit back with a dire prediction:

> I imagine fairly well that another VW factory in the US, provided that one more should still be set up there, does not necessarily have to be assigned to the South again. If co-determination isn't guaranteed in the first place, we as workers will hardly be able to vote in favor [of building another plant in the American South].[56]

This was an empty threat since southern states' willingness to aggressively incentivize new manufacturing was at least as important a factor in site selection as their aversion to unions. But by tying VW's expansion to the vote, Osterloh delivered a mirror image of Corker's earlier ultimatum.[57]

Union Substitution

Scholarly interest in Chattanooga centered on the role of the works council. American labor law stipulates that unionization is a necessary precondition for implementing "supplemental" employee representation schemes. Nonetheless, union staff presented affiliation with the works council as the central objective of the organizing campaign, while the union itself took a back seat. Despite widely circulated media reports to the contrary, at no point did any VW official go on record declaring support for the UAW, but the company repeatedly issued pro–works council statements. In the process, the union was undersold to members. Given the overriding emphasis on the works council, some workers concluded (incorrectly, but not unreasonably) that a works council might be achievable *absent* unionization. To fully explain the complexity of the works council's influence, we must examine the German labor relations system more closely before considering how it might translate to the United States.

German Industrial Relations

The German principle of codetermination provides an opening for worker participation in management decisions. The umbrella term *codetermination* refers to two distinct but interlocking systems: (a) employee representation on supervisory boards and (b) works councils. In Chattanooga, union and company officials only considered the works council prong of the codetermination system; it is unclear how employee representation might be achievable under U.S. labor law.

Under codetermination, workers are granted voting rights on supervisory boards of large firms.[58] In comparative industrial relations, the German model earns high marks for formally incorporating worker voice, granting employee representatives a direct say in the company's most vital decisions. Works councils are the primary mechanism for employees and managers to negotiate work arrangements and resolve disputes. In addition, works councils have broad regulatory oversight to ensure compliance with collective bargaining agreements and labor laws. Works councils must approve any changes regarding work rules and line speed before they can be implemented.

Ironically, employers initially encouraged works councils to restrict the influence of trade unions. The 1952 codetermination laws mandated that most decisions pass through the works councils; employers had hoped to bypass trade union influence by effectively creating their own company unions. Undeterred, trade unions have maintained a controlling influence over works councils. However, a strict division of labor remains between labor unions and works councils, with unions primarily focusing on "bread-and-butter" issues (such as wages and hours) and works councils addressing broader concerns, including working conditions and workplace organization. Today, trade unions have a close, codependent relationship with works councils that solidifies and maintains their mutual power while carving out distinct roles for each, with minimal overlap.[59]

To some extent, codetermination serves as a meaningful check on what U.S. labor law considers "management prerogatives." At the plant level, management is obliged to disclose economic and financial prospects, productivity metrics, and future hiring projections. At larger firms, works councils can inspect financial statements and obtain information about the sale or purchase of assets. Works councils also have input in hiring and firing decisions, including the right to veto terminations.[60]

The role of the state in German industrial relations is also significant. While conventional wisdom holds that codetermination consigns the state to the role of junior partner in what is ultimately a deal between employers and unions, the German model is an authentic tripartite alliance that bears the state's handprint at every level. In this sense, the German state is both a *facilitator of* and a *party to* labor agreements.[61]

Among the most vital aspects of the German industrial relations system is its resilience. The rules governing the selection of employee representatives have been one of the most hotly contested aspects of the codetermination system, yet have remained practically unchanged for sixty years. Unions and employers have attempted to tweak the rules at various points, yet apart from minor changes that do not fundamentally alter the balance of power, neither has succeeded.[62] The German system has stood the test of time because it allows stability and predictability, with 63 percent of the German workforce currently represented by German unions.

Codetermination with American Characteristics

Having examined the intricacies of the German system, we now explore its application in Chattanooga. Despite the firm's ambivalence, stalling, and backtracking, inside sources confirm that a vocal faction of senior management genuinely desired a works council, perhaps to maintain uniformity across their multinational operations. From the company's point of view, a works council would bring a measure of parity between Chattanooga and the rest of VW's far-flung global production network. Wall-to-wall works council representation might allow VW to achieve industrial peace. After all, as the German experience demonstrates, works councils are entirely compatible with a certain version of capitalism. This faction viewed the union as an unfortunate but necessary waystation on the road to codetermination, given the strictures of American labor law.[63]

U.S. labor law prohibits yellow-dog unions (those governed or controlled by a company). As conceived, the National Labor Relations Act (NLRA) assumed adversarial relations between employer and worker and positioned the state as the mediator, akin to a courtroom judge, of this natural antagonism. In this model, neither party can dominate the other, lest the agonistic system yield unfair results. The drafters of the NLRA aimed to ensure that unions would serve as genuine, democratic representatives of workers and prevent unsavory employers from installing management-friendly figureheads to undermine authentic member-controlled unions. Legal scholars broadly agree that works councils cannot be installed without an independent union organization, lest they run afoul of the NLRA's domination and interference clause. However, most legal experts agree that on its face, U.S. labor law does not ban works councils outright or deny them standing.[64] Once the necessary prerequisite of unionization had been cleared, a prospective works council could be bent and molded to fit U.S. legal codes.

While VW insists on extending the reach of its works councils to all plants, its works council model is flexible enough to adapt to the contingencies of local environments. Works council power can be titrated to match variance in industrial relations systems. As a result, GWC affiliates vary enormously in scope and purview depending on the country in which they are based.[65]

Communications from the company consistently stressed the works council while downplaying the union, which it presented as a little more than an obligatory prelude to the main event. Outside commentators weighed in with similar views, with no less an authority on the issue than Thomas Kochan declaring that a works council "could usher in the next-generation labor-management partnership" he had been advocating for decades.[66] Notably, his comments likewise avoided the thorny issue of unionization.

This section has shown that management embraced the works council as a form of "unionism-lite": a mode of employee representation based on partnership, not confrontation. Even though there is no legal basis to install a works council without a union, this illusory vision distracted from and compromised the union's efforts. The following section shows that the works council scheme proved enormously unpopular among workers, actively undermining the campaign.

Works Council

The works council stole the show in 2014, distracting from the overriding objective of unionization. Unfortunately, the union was often complicit in the sidetracking of its campaign. Conventional organizing strategy teaches that prospective members must come to see a union as the logical solution to their day-to-day problems. But with the works council's role and purview still vague and poorly defined, and with no meaningful precedent for codetermination in the United States, persuading members that a works council would do anything to remedy their day-to-day grievances was a challenge. The logical association between a works council and day-to-day issues on the shop floor was always murky.

When selling the works council to workers, union leaders echoed management in describing the works council as a leveler. IG Metall (IGM) international affairs director Horst Mund urged Chattanooga employees not to be the "odd man out" by rejecting the UAW, later noting:

> One of the reasons VW is arguably the most successful company in the world is that in every single one of their facilities, with the excep-

tion of Chattanooga to this point, they have employee representation.[67]

Similarly, in an editorial for the local paper, UAW regional director Casteel advocated for the union on the basis of maintaining uniform corporate governance: "Chattanooga should not be the only facility outside of [the works council] system. For the Chattanooga plant to be an outlier weakens its position inside the VW system."[68] For its part, the works council naturally sought expansion and self-perpetuation for reasons of self-interest as it believed that the company's success depended on achieving full representation across all of its international operations. As GWC chair Osterloh put it, "VW has only acquired its global strength because workers are tied to corporate decisions."[69]

For the UAW, collaborating with the company on the shared goal of building an "American-style" works council was a backdoor means to secure unionization (though chronologically, unionization would necessarily come first). The UAW also made a series of early statements promoting the works council, perhaps to appease management in its quixotic bid for voluntary recognition. In what can only be described as a gross overstatement, UAW president Bob King doubled over backward to present his union as ideologically compatible with the works council tradition, saying, "The German co-determination system . . . is completely consistent with the UAW's 21st-century model of unionism."[70]

However, the centerpiece of VW's peace offering had little appeal to workers, who saw it as part and parcel of the company's duplicity. Many felt that the works council was closely integrated with the company to the point that it could not serve as an independent arbiter of worker opinion or advocate effectively for their interests. These attitudes may have resulted partly from a weak grasp of comparative industrial relations. Although they were savvy and well informed in most ways, workers had only a vague and often misguided understanding of the works council's activities abroad, let alone how it might behave and integrate with a union in the United States. In sharp contrast to outside commentators, who often viewed the works council as a worthy goal in itself, workers saw it favorably only to the extent that it might pave the way for a union with genuine collective bargaining capacity.

A minority of workers had unrealistic expectations of the works council, which they imagined as an all-powerful entity that could deliver unionization via proclamation. But in general, workers were skeptical of the works council, which they saw as closely integrated with the company to the point that it could not be an effective arbiter of worker opinion. Some workers were unsure whether the works council would vociferously advocate on their behalf. For all the attention the works council received in the press, it had little relevance for most workers. This is understandable, given that neither the works council nor the UAW articulated a clear or consistent vision for how they might coexist in an American context (pending recognition). But the UAW organizing committee was not sufficiently prepared to dispel misconceptions about the works council within their midst, let alone correct the falsehoods spread by the opposition.

Organizers' critiques of the works council fell into three broad categories: (1) it would be ineffective, (2) it was a poor substitute for a union, and (3) the company was ingratiating itself with the works council as a way of avoiding unionization. Many workers were dismissive of the works council, which they saw as a bureaucratic fix that would do little to resolve everyday issues around the plant. One typical attitude:

> My co-workers don't give two s—— about a works council. Most don't even know the functions of a works council. What they care about are solutions to the problems we face daily, like line speeds, break times, and overloaded pitches. Those are things needing immediate action.[71]

Many workers claimed the works council was a powerless entity that lacked either the ability or the willingness to influence the unionization vote. In fact, workers were skeptical that the works council had any value at all in the absence of union recognition. One described his ambivalent feelings toward the works council:

> If the works council wasn't going to leverage card-check recognition, then I didn't see any reason to try to form a works council without major concessions from the company. When I got involved in this, I had never heard of the works council, and it is just icing on the cake of a real UAW local, in my opinion.[72]

Another worker, similarly, made it clear that his loyalties were aligned with the UAW, not with the works council:

> I don't remember joining a works council organizing drive. I joined this drive to organize a union. From what I've seen of the works council, they make a lot of promises to placate us until they get their way. I am more apt to say, let's just go straight UAW and screw the works council. I know at least with the UAW, someone will actually give a s—— about my concerns, problems, grievances, or harassment.[73]

Some workers considered the works council a shill for VW, a misconception that gained traction when the works council refused to openly advocate for the UAW. These sentiments were only confirmed when workers noted that the company granted the works council much better access to workers than the union. For example, the works council received permission to station a full-time representative inside the plant,[74] and while the union had a significant presence in the plant, it was accommodated less consistently. One worker went so far as to suggest that the works council would mitigate against unionization:

> The works council might be a double-edged sword. . . . The places where things like that do exist, it's just so the company maintains control and gives you a sense that you have a say when you don't. And that would be disastrous.[75]

Finally, there were deep concerns that the works council (and its affiliated union) would not be sufficiently independent of VW, given the extensive commingling of union and works council that is common in Germany:

> You're basically—you're paying dues to a department in the very company you're working for. When they can go in there and have these pat-on-the-back feel-good meetings with the company and the workers doesn't even know about it—that's ridiculous. That's wrong.[76]

Eventually, workers came to view the works council as inseparable from the company and its duplicity. One put it succinctly:

> VW loved it so much because they could use it as their personnel department; they could use it to run interference between them and the workers.[77]

As many scholars have observed, international solidarity requires coordination between powerful entities and a degree of buy-in from the membership.[78] The UAW fell short on both counts, receiving only nominal support from the works council and failing to persuade workers of the importance of its international alliances. As bold and ambitious as its works council proposal was, it never concretized into a workable plan and was undersold to the rank and file. Though workers understood that foreign parties could aid their drive, they were not deeply committed to the international aspect of the campaign, and the UAW made no real attempt to convince them otherwise. Without an educational component, the UAW's transnationalism was less an executable game plan than a notional vision statement.

Deflection and Distancing

The instability and volatility of VW's corporate culture presented a final organizing challenge. There is a vast literature on the transferability of management practices overseas and the barriers to cross-cultural communication. But practical problems such as language barriers and cultural differences do far less to inhibit interunion coordination than divergent national labor market institutions.[79] At VW, the company's cultural norms were challenged by a legal apparatus that offered no practical outlet for a works council and an American tradition of corporate anti-unionism that proved all too pervasive.

VW in Chattanooga is a test case for transferring European-style labor relations to the United States. Unlike typical transnational corporations (TNCs), VW has tried to preserve its distinct corporate culture in its foreign operations. In keeping with this practice, it attempted to export some version of its works council to all foreign plants, even in countries with challenging labor laws. The differences in auto industry labor conditions between Germany and the United States are considerable. For example, German workers have traditionally enjoyed extensive protections against layoffs, which are contractual and cannot be terminated by either side. Given the near total absence of such protections in the United States, many observers were eager to learn whether VW's elevated labor standards could survive the company's perilous transatlantic voyage.

The past decades have seen several European-based TNCs with socially responsible employment practices establishing operations on U.S. soil. However, rather than maintaining consistent business practices throughout their global supply chain, these firms often abandon their corporate culture at the customs booth and neglect the most rudimentary labor rights when dealing with American workers. According to Lance Compa, who has studied many such firms, this "when in Rome" approach is not uncommon: "Foreign companies argue 'the U.S. is different' and mutate their labor relations policies into U.S.-management style campaigns against workers organizing efforts."[80] Likewise, a comprehensive study of German firms' behavior in the United States revealed that German firms consistently "aspire to adopt the local practices of workplace organization and employment relations. These aspirations represent their attempt to fit into the local competitive environment—in other words, a process of adapting to the local, regionally specific institutional context."[81]

Consistent with these findings, there appeared to be a disconnect between the company's mode of operation at its sixty-one overseas plants and in Chattanooga—its sole facility on American soil. Indeed, the Chattanooga plant bore a closer resemblance to legacy American automaker factories than to

VW operations globally. Although VW had a reputation for preserving its distinct corporate culture even as it expanded to foreign plants in South America and eastern Europe, it was quick to assimilate in the United States, changing its colors to match the local environment.[82] A UAW staffer noted:

> [Managers] meet with a replacement coming in every two years. And then they say, "This is not VW. This is Ford, this is GM, this is Chrysler, and this is Toyota." It just depends where you land in the plant. They destroyed that company culture. Their own engineers would go home and say, "This isn't VW."[83]

When describing the plant's internal culture, some workers used a different national reference point: Japan. American companies had partially adopted the Japanese model long ago, so the mixed metaphors are all too appropriate. But, according to some workers, the company's conduct was so far removed from the German model that it more closely resembled the Japanese. Given that no other auto-manufacturing facility exists in the Chattanooga area, VW persuaded the city to erect the VW Academy training facility to train an otherwise underprepared workforce. Yet, although the academy inculcated workers in German-style production techniques, these proved almost useless once workers graduated and set foot on the factory floor. One worker observed this disconnect firsthand:

> The whole plant was supposed to be set up on the premise that they were the German way of making cars. And when you went through training, when you went to the academy, that's what they told you. The German way, the smart way, systematic, you do it right every time, attention to detail. When you get to the floor, there's no Germany down there. You're in Japan. You walk down there, and you're in Japan.[84]

In one telling, anti-union American managers were manipulated by their pro-union German overlords. Corker first hinted at this supposed rift between American and German management, informing Reuters that U.S. executives had been "forced" to disclose the company's talks with the UAW and that a German board member had imposed the open letter to employees on plant-level leadership. Whatever the merits of this claim, Corker believed that by exposing "rifts" within the firm, he could foment internal tension that might be exploited later. This was either baseless speculation or a calculated strategy to undermine the firm by dispensing bits of (uncorroborated) corporate intrigue.

In terms of management philosophy, VW presented its Chattanooga plant as a polyglot combining the best of three worlds: Germany, the United States, and Japan.[85] In a premonition of things to come, VW sought to build a management team that combined homegrown German experience with American-sourced auto industry veterans, themselves often borrowed from Japanese transplants. The first plant president, Frank Fischer, a veteran of many different VW plants across the globe, noted that the plant design incorporated elements "from 61 factories, from Russia and India."[86] The Japanese influence was embodied by Don Jackson, among the plant's most vocal anti-union voices, who was hired from a Toyota plant and served as Chattanooga's HR director for its first year.[87] In a nod to local mores, Fischer also mentioned that the plant was still distinctly American despite its international composition.[88] As time passed, the plant would hire more American managers, chiefly from nonunion Japanese transplants (especially Toyota's Kentucky operation and Nissan's plant in the Nashville hinterlands).

Expectations that VW might retain its company culture when doing business abroad were not entirely unfounded. Among multinationals, VW is unusual in its long-standing reputation for building tightly integrated global operations where a culture of deference reigns supreme and German meddling in everyday affairs is an accepted part of company culture. Although all car manufacturers are, in some ways, products and reflections of the economic context in which they are based, VW stands out for having especially deep historical ties to the German economic system and embeddedness within German institutions. As late as 2010, VW was still more a multinational than a fully realized transnational company, with peripheral manufacturing sites wholly dependent on the German metropole.[89] In contrast to the iconic River Rouge and Hamtramck factory complexes of Ford and Chrysler, which had long since passed their heyday, VW's enormous Wolfsburg factory was still its standard-bearer. Relative to its key competitors, VW's transplants were undercapitalized and generated uneven returns on investment. Its first foray into the East German frontier was fraught with missteps and false starts. In 1992, VW set up a plant in Mosel, but operations "were stalled and frustrated from the beginning" as the plant "lacked a consistent strategy and commitment on the part of corporate management."[90] Rather than "NUMMI in Saxony," as the plant was inappropriately nicknamed, the operation was a hybridized crossbreed with German and Japanese production characteristics that competed for influence without ever achieving a stable equilibrium.[91] With its heady and sometimes incoherent mix of HR management strategies, the Chattanooga plant threatened to fall victim to the same problems that had doomed Mosel. At times in Chattanooga, evidence of the growing disconnect between Germans and Americans played out on the shop floor. For example, the plant had issues with quality and volume during its early years. Though VW had a reputation for running its international plants with a tight grip, local plant managers almost immediately received the go-ahead from superiors to address quality issues in-house without first consulting top-level German executives.[92] Such conflicting directives and false starts became characteristic of the chaos that played out on the shop floor, with direct implications for the union drive.

Yet from the capital's standpoint, offshoring creates a physical separation between a firm's central base of operations and its peripheral investments. Although firms attempt to export the brand's essential identifying attributes to foreign production sites, globalization often dilutes company culture—sometimes beyond recognition. Capital flight may be motivated in the first instance by the need for proximity to consumer markets, but labor market considerations are always front of mind for high-sunk-cost items like automobiles. Capital moves to lower production costs, with labor consistently the most significant line item on the balance sheet.[93] Rather than maintaining consistent labor practices throughout their global supply chain, these firms typically reduce standards to maximize profits. Varieties of capitalism[94] compel firms to mutate as they cross national borders, changing their colors to match the local environment.

The writers Thomas Dunfee and Timothy Fort have proposed an intriguing metaphor for modeling the process of context-dependent cultural change: the "corporate chameleon." As they write, these creatures

> follow local norms and customs, which will generally be considered to dominate or trump home office policies or any universal standards. They do not merely respect local customs and traditions; they fully adopt and internalize them in their foreign operations. Thus . . . there may be substantial variances in norms and attitudes toward certain ethical issues. The norms of the local environments in which the firm operates become part of the overall value system of the firm.[95]

Dunfee and Fort contrast this model with the more conventional "nationalist" firm, which traipses the globe as an ambassador of the home country, exporting native traditions to its far-flung theater of operations without regard for local customs.

In short, VW presented itself as something of a cultural hybrid. If it imagined itself as a corporate chameleon, it was never a terribly convincing one. Workers came to believe the company was talking out of both sides of its mouth. The company hired known union busters as midlevel managers—including Don Jackson, whose résumé included a successful defeat of a union at Toyota—while at the same time, it remained a signatory to one of the most ambitious and far-reaching global labor agreements. VW's two-faced conduct in Chattanooga dealt the firm significant reputational harm, which was only partially mitigated by the subsequent emergence of a more serious emissions scandal later that year. It is not difficult to understand the corporate chameleon's enduring appeal. Adaptability is surely a desirable attribute in the global economy. But the corporate chameleon is not only quick to adapt; it is also a master of disguise. There is something fundamentally deceptive about its context dependency. Sometimes, our metaphors reveal more than we intend. In practice, a firm's command structure ensures that the adjustment to the local environs can be partial at best. This leads to outcomes that may read as indecisiveness, organizational incoherence, or even internal systems failure.[96] Unlike our chameleon, which transforms gradually as it plods from desert to sea, the firm must simultaneously inhabit multiple environments and maintain all of its operational modalities concurrently. At this point, it becomes difficult to distinguish the original firm from its cheap imitation. Thus, our firm risks becoming a vulgar postmodernist with no primary referent, only unstable meanings. There is no true essence behind this animal's coloration.

Conclusion

This chapter has shown that the UAW's organizing campaign was stymied by a spate of goodwill, a series of promises (many disingenuous), the company's focus on implementing a works council rather than supporting unionization, stalling and ambivalence, and opposition from politicians and anti-union groups, which intensified when VW delayed expansion decisions contingent on the vote results. Yet, as it licked its wounds, the union pinned the blame squarely on political interference. Corker and Haslam made for easy villains with their theatrical flourishes, but reality is not a morality play. Blaming the opposition did not signal a path forward, as the political climate was perhaps the one constant amidst a background of uncertainty. As we will see in the following chapter, the union also bore some responsibility for its own defeat.

3

Solidarity Subdued

Why the UAW Lost in 2014

The previous chapter identified eight challenges the union encountered as it tried to organize the company in 2014. As serious as these obstacles were, they were not unexpected. Each was a recurring theme across thirty years of failed organizing drives at auto transplants. Returning to the strategic capacity framework, this chapter shows that with better institutional memory (knowledge), the UAW could have gleaned lessons from previous failures to prepare itself more adequately for its difficulties. Relatedly, a more vigorous campaign with stronger motivation would have positioned the UAW to overcome whatever additional challenges it might confront.

Additionally, more adaptability and faster feedback mechanisms might have encouraged learning, permitting the union to prevail even as the terrain shifted beneath its feet. While the UAW earns higher marks for innovation, it squandered its boundless creativity because it lacked easy access to good information: its bold plans and grand schemes had no material grounding. In short, the UAW scores poorly on three out of four strategic capacity measures, downgrading its overall performance despite some intriguingly innovative ideas.

Knowledge

As the UAW set its sights on Chattanooga, it lacked complete and accurate knowledge, which should have been informed by its past efforts to organize transplants. The UAW should have been able to summon the relevant information in applied contexts, make it accessible (salient), and clearly and efficiently disseminate this knowledge among affected parties. It fell short on all these counts. The union suffered amnesia when it should have experienced déjà vu. Though the organization had a vast stockpile of knowledge that should have allowed it to decode VW's mixed signals, it neglected to make this knowledge available for reasons that overlap with its lack of motivation and misdirected innovation.

Since the 1990s, the UAW had repeatedly tried and failed to organize foreign automakers, initiating a new campaign roughly every three years. Positioned against the *longue durée*, the 2014 effort at VW was the UAW's seventh attempt to organize a foreign assembly plant, and the win in 2024 was its tenth. The UAW's approach to the Chattanooga campaign was hampered by its failure to learn from past experiences with organizing foreign transplants. There were striking parallels between these initiatives, and the failure to recognize them was in no small way responsible for the UAW's defeats.

Despite VW's initial openness to a works council, the UAW's efforts mirrored previous unsuccessful campaigns at other auto transplants. Virtually all the obstacles the UAW faced in Chattanooga were carbon copies of tactics deployed against it at other transplants in earlier elections. Generally speaking, VW's site selection process, supplier network, workplace governance, management philosophy, and post-2014 anti-union conduct are consistent with those of other foreign transplants in the South. One differentiating factor stands out: VW broke with the dominant pattern in its labor relations posture from 2013 through 2014, especially its early embrace of the works council. Otherwise, in its broad contours, the UAW's experience with VW is similar to that of failed unionization drives at Nissan in Smyrna, Tennessee; Honda in Maryville, Ohio; and Mercedes in Vance, Alabama, as well as aborted campaigns at Toyota in Georgetown, Kentucky, and BMW in Spartanburg, South Carolina. The failure to recognize these patterns was in no small way responsible for the UAW 's defeat.

Comparative studies of foreign transplants reveal only slight variations on a common playbook. Companies are lured to the United States with the promise of generous incentives, forged on the basis of personal relationships cultivated between politicians and senior executives, and further buttressed by the promise (whether implicit or stated) of a low-wage ceiling and a union-free environment. Once the deal is signed, the company is treated to an extended honeymoon during which local press and politicians regale the newcomer with unfettered praise. In all these respects, the VW saga was typical. The extraordinary aspects of the campaign largely served as distractions from what was, at its core, a fairly conventional struggle between the autoworkers and a foreign auto transplant.

While contemporaneous accounts stressed the "exceptional" aspects of the UAW's fight in Chattanooga, in its key elements, the struggle to organize VW was typical of unionization campaigns among auto transplants and more broadly reflected the challenges unions face in the contemporary United States. If journalism trades in a constant stream of superlatives that serves as grist for its breaking news feed, social science is charged with the less glamorous but arguably more essential task of seeking out generalizable patterns across time and space.[1] On the latter count, VW did not disappoint. To be sure, the fight had all the dramatic elements that helped propel it to the top of the news cycle—outsize personalities, elaborate staging, transnational power brokering, and heightened rhetoric—but behind the scenes, the terms of struggle were more mundane, even familiar. Dispense with the protracted works council debate, tone down the hyperbole, and the failed organizing drive in Chattanooga looks suspiciously similar to UAW's other entanglements with transplants.

In addition to these broad parallels, specific lessons might have been learned from past attempts to organize transplants. For example, the UAW was blindsided by a duplicitous neutrality pledge at Honda in 1982, confronted with a rumored plant expansion amid a vote at Nissan in 1989, and schooled on the limits of transnational alliances through its work with IGM at Daimler in 2009.[2] As it turned out, if there was a secret formula that might allow the UAW to break the spell and win over transplants, it was to be found not in the sleek boardrooms of VW's global headquarters but in Midtown Detroit, inside a squat midcentury-modern public university library that houses the union's official archive. In sum, apart from some early deviations, the company's behavior conformed to the dominant pattern of auto transplants faced with the prospect of unionization.

Motivation

Even in its best moments, the union ran an anemic campaign undermined by a series of problems, the cumulative effect of which proved debilitating, compounding and metastasizing into spirals of self-destruction. The UAW's strategic capacity was undermined by a near-complete disregard for widely held guideposts that structure comprehensive campaigns and sometimes by more fundamental incompetence. Even more than external factors (discussed in the previous chapter), these internal missteps prevented the UAW from achieving a viable organization and cemented its defeat.

While there is no consensus on the particular attributes of a union campaign that predict success, the literature suggests some general standards. Kate Bronfenbrenner and Robert Hickey find that positive NLRB election

outcomes correlate strongly with the use of multiple "comprehensive organizing tactics."[3] No single tactic emerges as a silver bullet, but their combined
use yields good results. In another text, Bronfenbrenner and Tom Juravich
write, "Union success in certification election and first-contract campaigns
depends on using an aggressive grassroots rank-and-file strategy focused on
building a union and acting like a union from the very beginning of the campaign."[4] It would be difficult to argue that the UAW acted like a union during
the lead-up to the 2014 election. Successful campaigns tend to maximize personal connections, but most workers had little to no contact with union
representatives or supporters. Nor did the union use the intensive rank-and-
file approach these studies suggest is most effective. In fact, of the ten "comprehensive organizing tactics" recommended by Bronfenbrenner and Hickey, it appears the UAW incorporated at most two: "active and representative
rank-and-file organizing committees" and "active participation of member
volunteer organizers." Strategic miscalculations led the UAW to run an un-
inspired campaign nearly devoid of energy and enthusiasm. Rather than building rank-and-file support through conventional means, it relied heavily on
the power of an unproven alliance with the works council. But this was a risky
strategy, since U.S. labor law denies works councils standing at nonunion
employers and makes unionization a virtual precondition for works council
representation.[5] Despite these stiff odds, in the lead-up to the election, the
UAW seemed to subscribe to the deus ex machina theory of labor organizing,
wherein an outside power intervenes to deliver the union without real mobilization on the ground. By placing all its stock on its backroom dealings with
the works council, the UAW left its front-of-house operations diminished, un-
supported, and vulnerable to outside forces. Afraid to offend the employer or
the works council, the union led a quiet, low-key, and, by the lead organizer's
admission, low-budget campaign. Perhaps even more seriously, it suffered
from a disproportionate underrepresentation of affected groups, including
people of color and unskilled workers, which led to structural weaknesses
that only worsened over time. Excluded from high-level meetings, workers felt
the UAW was colluding with the company. Equally important, in an effort to
keep their restless rank and file from disturbing the delicate alliance they had
built, organizers resorted to authoritarian tactics, stifling dissent and fostering
internal division, which ultimately led to a demoralized organizing committee.[6]

Although the UAW had considerable organizational capacity at the international level, its resources were not deployed readily or effectively in
Chattanooga. Motivation is a measure not of an organization's capabilities
but of the degree to which the organization manages to mobilize these capabilities at a given historical juncture. According to data from the independent researcher Chris Bohner, the UAW is the wealthiest union per capita,

with $3,060 in net assets for every member it represents.[7] But as of 2020 (the closest year for which data is available), the UAW spent only 6 percent of its annual budget. For comparison, the constitution of the Service Employees International Union (SEIU) requires that the organization dedicate at least 20 percent of its resources to organizing.[8]

The UAW's efforts departed from the ideal organizing model document-ed in the literature in many respects. At the height of the campaign, the union had a maximum of four full-time staff organizers committed to the drive, or roughly one organizer for every four hundred eligible workers, and rank-and-file engagement was rarely a priority. Even at its peak, the UAW's campaign can only be described as anemic. In the words of one participant:

> They didn't commit the resources. They didn't organize. They didn't run a traditional campaign. They relied solely on the good graces of the company. They put all their cards in this idea that backroom deals could result in this worker council.[9]

Militancy

As later exposed, the pared-down campaign was no mere strategic calcula-tion but the pound of flesh VW had demanded in exchange for its neutral-ity pledge. Unbeknownst to workers, the UAW had signed away its most powerful organizing tools in exchange for modest concessions. As it turned out, the neutrality pact was conditioned on a promise from the UAW to scale back its organizing efforts and avoid confrontational tactics. Under the terms of this agreement, the UAW agreed to forgo house visits, refrain from picketing, and dispense with recognition strikes. In return, it would be granted access to contact information for all plant employees, in-plant office space, and permission to display materials in nonwork areas.

Further, VW granted the union permission to conduct pro-UAW captive audience meetings at the factory and agreed to provide staffers with broad (although not unrestricted) in-plant access to workers.[10] Finally, VW and the UAW agreed to align and coordinate their public relations messages. In prac-tical terms, the agreement's impact was felt mainly through the exclusive access provisions. But a passage buried deep in the document attracted far more at-tention:

> The parties recognize and agree that any such negotiations for an ini-tial collective bargaining agreement and any future agreements shall be guided by the following considerations: a) maintaining the high-est standards of quality and productivity, b) maintaining and, where possible, *enhancing the cost advantages and other competitive advan-*

*tages that VWGOA [VW Group of America] enjoys relative to its com-
petitors in the US and North America, including but not limited to legacy automobile manufacturers.*[11]

Though the lack of specifics left some room for interpretation, this passage
promised the UAW would not challenge the low wages that had attracted VW
to Chattanooga in the first place. Specifically, since starting pay for VW Chattanooga was at least eight dollars per hour less than starting pay at unionized
("legacy") Big Three plants, many saw this language as thinly veiled code for a
promise that Chattanooga would continue to undercut Detroit. Given that
the regional pay differential was presumably among the UAW's deepest concerns, this raised serious questions about the union's strategic position. For
labor expert Jefferson Cowie, the language on "maintaining cost advantages"
seriously undermined the UAW's cause, suggesting that it "didn't have a lot to
offer besides the works council. If they're not going to bargain hard for wages,
then what are the workers going to get?"[12] When confronted with these ideas
and asked to explain the damning language on "competitiveness," UAW regional director Casteel offered little comfort, merely noting that "pay was not
the central focus of the Chattanooga organizing campaign."[13] Although making
wages competitive with Detroit would seem to have been high on the UAW's
agenda, the issue never got much attention under the terms of the agreement.[14]
Given that wages were deemphasized, as one commentator wrote, "Ultimately,
all the UAW had to sell were soft concepts such as worker democracy."[15]

Still, in pandering to the company, the UAW exceeded what its deal required. It was reluctant to take a firm stance against everyday mistreatment
for fear of jeopardizing its nascent partnership with the company. Since its
opponents had already cast the UAW in the image of 1950s-era thuggery,[16]
staff felt the need to disprove this stereotype by (over)emphasizing their civil-
ity. They believed that taking the high road and presenting a sharp contrast
to the opposition's aggression could persuade conflict-averse workers to join
their cause. Staff downplayed whatever militancy the UAW had left in its
quiver and recast themselves in the mold of southern good old boys. Typical
(nonactivist) workers prefer industrial peace to the chaos and uncertainty
of a contested election, even if militancy might "deliver the goods."[17] Trading
on stereotypes of southern humility, one staffer said:

> I would say—and I hope that I am wrong—that what the southerners
> would like is a very hierarchical, authoritarian, management-friend-
> ly union.[18]

But the UAW went well beyond avoiding combative tactics and shied
away from antagonizing the company at all. This ended up costing it support

even among its committed backers, one of whom expressed disdain for the union's approach:

> [The UAW] was too soft about it. When VW did step out of line or when there was intimidation or harassment in the plant, they were too friendly with the company about it. The skeptical, hardcore pro-union of us were getting disillusioned . . . not the kind of union I really wanted.[19]

Because the UAW had so much confidence in VW's professed neutrality, it banned the expression of any negative sentiment about the company. As one worker noted:

> They never criticized VW. Discouraged us from criticizing them. . . . I felt that was a VW tactic. VW—that was their way of controlling the situation—was to play nice.[20]

Another suggested that the UAW's desperation to organize the plant de-fanged its campaign:

> They were willing to throw VW a bone and give VW whatever they wanted just to get the UAW shingle hung outside of that place. They were so desperate for a foothold in the South that they thought they could get in here and strong-arm everybody into organizing.[21]

Even as the campaign grew more heated, rather than responding in kind, the UAW tried to ratchet down the antagonism with a muted reaction to the nascent opposition. Some workers would have preferred a more confrontational approach:

> It almost feels like the UAW is afraid of any kind of direct action, because they're afraid it will jeopardize whatever they're working on with management.[22]

UAW regional director Casteel reiterated his desire for the drive to stay "low-key, low-budget" and presented the union's minimal presence on the ground as evidence that the campaign was a "homegrown deal."[23] Similarly, UAW president King told reporters at the Detroit auto show that the UAW deliberately slowed its organizing push in the run-up to the vote: "We consciously tried to de-escalate all the public stuff."[24] Cynics, including Chattanooga community activists who had long advocated a more visible campaign, observed that there had never been much in the way of "public stuff."[25] Instead, what had always been a quiet campaign was driven even further underground.

Communication and Transparency

Even if the UAW had good intelligence on its adversary, communication break-downs prevented it from sharing this information with members. As a result, prospective members felt shut out from the organizing process, and even adherents were marginalized from the very organization they sought to build. In the worst moments, this led to mistrust, finger-pointing, and the eventual self-destruction of the organizing committee.

Though there are real challenges inherent in a drive involving 1,700 workers, the UAW organizing staff consistently struggled to deliver timely, accurate, and clear information to its worker-organizers throughout the campaign. One of the biggest causes of frustration was the absence of reliable information about the status of the card-check agreement. For a year, workers were left to speculate about whether the company would consent to card check, whether majority support had been achieved, and, after card check was abandoned, if and when the election would occur. Several members alleged that the union lost credibility by compelling workers to submit to the ritual of card check, only to sit on the cards for close to a year, then abandon card check in favor of an election. The union was not to blame for the aborted card-check procedure, but it failed to offer a convincing and transparent explanation for the abrupt change in strategy. This communication failure added to the sense that the union was a third party, distinct from the workers it sought to represent.

The UAW also struggled with messaging. Injuries and scheduling were the most critical concerns among the workers in my sample. Yet the UAW did not emphasize these themes. The opposition's consistent, overwhelming, and highly targeted communication rendered the union's lack of attention to messaging and willful neglect of certain hot-button topics particularly glaring. As noted previously, the opposition drew on generous outside funding to blanket a major highway with display ads.[26] The union's response to these initiatives was tempered, with UAW regional director Casteel dismissing the showy advertising as a "waste of money" and refusing to respond in kind.[27]

The union's organizing committee was an amorphous and evolving body from the start, but UAW staffers struggled to maintain coordination and unity of purpose within this group.[28] Adding to the difficulty, staffers seemed unable to develop a consistent and predictable method for calling meetings, relying on a mixture of text messages, in-person contact, and back-channel word of mouth. Eventually, these problems led to charges of favoritism, as certain organizing committee members felt excluded from meetings simply because they were not privy to haphazard and inconsistent modes of communication. At times, staff did not communicate at all: long periods passed during which even core council members heard only radio silence from the

UAW. Poor communication also impeded workers' efforts to initiate organizing activities. One worker described silences of seemingly endless duration:

> There were stretches of time, sometimes up to six months, between any kind of communication or press release from the UAW. I was over there [at the union office] sometimes two or three times a week, and we were begging, "Let us put out flyers." They could just be informational. Because the people who were against the union were sure out there talking about it. You've only got one source of information, and there's a vacuum in there; that's what's going to fill it up.[29]

In the absence of clear communication from above, workers resorted to back channels of their own. Workers became especially incensed when given the impression that the union was engaged in machinations to which they had no access: "Some kind of behind their backs, sneak something strategy."[30]

Since the UAW neglected to maintain open lines of communication, a perception arose that the union was dragging its feet. Workers came to feel as if they were in limbo, and this had a predictable de-energizing effect on the drive. The UAW may have missed a window of opportunity by deciding not to hold an election in 2013, as multiple workers reported that morale in the plant had reached a nadir by 2014, demoralizing potential supporters. Summing up this experience, one worker remembered:

> We sat at a meeting with the regional director of the UAW downtown. We weren't complaining about the process; we just wanted to know what was going on. The only answer we get to this is that we need to be patient.[31]

Deprived of timely, reliable information, workers subjected the company's positions to scrutiny and independent assessment rather than taking its public declarations at face value. They continued to probe the recesses of the information vacuum for any clues as to the company's true stance. In the UAW's defense, the task of international coordination with a foreign company on a constantly shifting playing field is complex and unenviable. But the UAW could have done a better job of maintaining transparency, providing regular status updates, and improving the sense of buy-in, even during periods of uncertainty.

House Visits

The literature on union strategy suggests that one-on-one "organizing conversations" are most impactful when they occur in workers' homes.[32] However, the UAW had relinquished its right to conduct house visits as part of its neutrality agreement. Some workers disputed the notion that house visits would have made a difference, especially in the specific organizing context of the South. In contrast to northern cities, where passersby regularly traverse high-density neighborhoods, Chattanoogan residential communities, they argued, tend to be insular and cloistered. A door knock from an uninvited stranger would be highly unusual and potentially perceived as dangerous. In this respect, notions of privacy and territoriality often take precedence over southern hospitality. As one worker expressed:

> No . . . there weren't any house visits because they felt that people would feel intimidated. In some parts of the country, maybe, but here people in Chattanooga don't want you coming to their house. A lot of times, they don't want people they know coming to their house, much less people they don't know.[33]

It may be true that geography and culture mitigate against house visits, but there are important exceptions. Three years after its defeat in Chattanooga, the UAW employed extensive canvassing in its campaign against Nissan, and its door-to-door approach was well-received, though that campaign was also unsuccessful.[34] Much earlier, house-to-house canvases were a component of Operation Dixie.[35] In a different context, the Student Nonviolent Coordinating Committee's Freedom Summer get-out-the-vote effort depended almost entirely on house visits. The issue of door-to-door campaigning in the South is a complex matter.

Community Support

Recent scholarship has emphasized the pivotal role of labor-community alliances in generating successful organizing outcomes.[36] This holds true even when other aspects of a union campaign are lacking; as Marissa Brookes writes, a union's coalitional power can compensate for weak institutional or associational power.[37] But union-community coalitions are most effective when they are deep, long-term, and reciprocal—rather than superficial and fleeting.[38] Community coalitions are especially important because they help unions win and create preconditions for future success. Conversely, unions that neglect to build coalitions miss a critical opportunity, as they will be compelled to start all future campaigns from scratch. Coalitions have also been celebrated for their ability to forge long-term alliances beyond the bounds of a particular campaign.[39]

On this count, the UAW's Chattanooga drive fell well short: apart from a few half-hearted gestures, the UAW never established meaningful connections with community groups. In recent years, Chattanooga has suffered from a lack of neighborhood organizing, so the UAW faced challenges locating suitable allies. But even though Chattanooga did not have much local infrastructure in the first place, the UAW deprioritized linking up with what few groups existed. It coordinated one-off service days with a food pantry and a local church, but brushed off the one community organization that attempted to engage it (Chattanooga Organized for Action).[40] Though it may have been unrealistic to expect the UAW to spearhead a major community mobilization, its near-complete neglect of local groups contributed to a sense among union opponents that the UAW was an outside force with no connection to Chattanooga. To the extent that the opposition was locally based, it could persuasively present the UAW as alien.

Even when the community was eager and ready to participate, the UAW failed to engage. The founder of a local tenants' rights group reported that repeated requests to draw the UAW toward meaningful community engagement had gone unheeded:

> We begged them to give us something to do to support the unionization effort. We had multiple meetings with [the organizing director] in which we were like, "Please." The community needs to be involved. The community on the other side is being involved—the anti-union forces are reaching out to community members.[41]

When this failed, community groups took it upon themselves to stage a series of pro-UAW events, hoping to generate favorable press coverage. These included a prayer vigil and a "Billionaires for Wealthfare" culture-jamming pseudoprotest. The UAW's response was tentative and standoffish, and the leadership preferred more traditional forms of support.

Race

Workers were recruited to the organizing committee haphazardly, without regard to racial balance. Historically, the UAW's record on building multiracial coalitions can generously be described as checkered. The 1941 Ford strikes, in which hundreds of Black people were recruited as strikebreakers, and sporadic subsequent anti-integration hate strikes marked particular low points, but recent scholarship has highlighted the centrality of Black people in building the CIO and revealed the surprising prevalence of interracial unions in the prewar South, even against the backdrop of segregation.[42] In their boldest moments, Black activists in the CIO actively challenged workplace segregation, laying the groundwork for the civil rights movement. With their unions serving as living examples of successfully integrated institutions,

Black workers could also push for change through the contractual apparatus: CIO unions secured agreements containing strong antidiscrimination language and enshrined antiracist provisions such as seniority-based layoffs.[43]

But interracial unionism did not succeed in all workplaces. Black and white cooperation flourished in environments like packing houses where different races performed similar tasks side by side. In contrast, environments beset by strict racialized occupational segregation gave rise to racialized patterns of organizing and resisted the cross-racial solidarity necessary for industrial unionism to flourish.[44] Southern white workers used unions to shelter themselves against job competition from Black people in the most egregious cases.[45]

Support for unionization is consistently higher among Black workers than their white counterparts, and Black workers are overrepresented as union members relative to their population size, both in the South and elsewhere.[46] Wary of this dynamic, auto transplants assiduously avoided constructing plants in areas with a large Black population, including urban centers and the Black Belt, preferring exurban greenfield sites where the available workforce is majority white. Some transplants have been explicit about their racial preferences. For example, Nissan selected a site in then-rural Smyrna partly based on the "homogeneity" of the area's residents, observing that "people are of the same race and have the same values and backgrounds."[47] Beyond this, many Japanese automakers were reluctant to hire Black people until antidiscrimination court orders forced their hand in the 1980s.

VW did not go on record with similarly racist views, but its site selection may have been influenced by similar criteria. The organizing drive at VW struggled to build meaningful cross-racial alliances, given the history of segregation and deep-seated prejudice in the region. The racial composition of the plant skewed white, with Black workers constituting perhaps 10 percent of the workforce and Latinx workers barely represented at all.[48] This made the plant far whiter than either the city, with 33 percent Black residents, or the surrounding Hamilton County, with 21 percent Black residents.[49] Several Black workers served on the organizing committee, but the de facto leadership was primarily white, with Black workers an active presence relegated to secondary roles. The union's public face was almost always white in media appearances and other events. Black workers' sympathy for the union effort was especially important given that they were overrepresented in logistics and the body shop, areas of the plant that were more anti-union. According to one white union supporter,

> Black workers, for the most part, were like "Of course!" I felt that there was a division racially in the plant about the areas that they worked in. I felt like logistics and body shop—the Blacks that worked in the plant—that's where they put them. Body shop was one of the worst

areas for intimidation from management, but we had a lot of support from them.[50]

Skill

The UAW's first contacts in the plant were skilled machinists, who were consistently among the most active workers in the drive, though they constituted only about 11 percent of the total workforce. Even as the UAW prepared for a plant-wide vote, it had early success recruiting machinists. UAW staffers initially hoped that the strong support from machinists would galvanize production workers. But workers' response to the prospect of a machinist vanguard was shaped by the fact that machinists had several layers of privilege that distanced them from the plight of lowly "shop rats." For one thing, machinists operated at a physical remove from ordinary production workers: whereas production workers were confined to their respective pitches throughout their shift (excepting breaks), machinists had greater mobility. Their freedom of movement naturally positioned them as potentially effective organizers: one machinist recalled that he was free to move throughout the plant throughout much of the day, even for non-work-related purposes.

Strong support for the union among maintenance workers can be partly explained by the fact that they were much more likely to have prior union experience; many were recruited from the recently shuttered Saturn plant, a former UAW stronghold located approximately one hundred miles away in suburban Nashville. Though there was once a glut of former production workers from Saturn, many returned once the Saturn plant rebooted as a GM facility.

However, machinists were often viewed as elitist coffee-drinking slackers because during a typical day, they experienced more downtime, barring some catastrophic equipment failure. Claims of privilege were borne out in demographic differences: machinists as a group were more highly educated. They had more previous manufacturing experience than production workers, many of whom had never attended college or entered a mass-production facility. Therefore, despite their mobility, machinists may not have had close contact with production workers and were not necessarily able to shape their attitudes. This divide was unfortunate, given that machinists played a disproportionately large role in the organizing committee: of roughly a dozen highly active worker-organizers, machinists held three spots, far in excess of their representation in the actual plant.[51] Considering that workers placed such a high premium on trust, it may have been difficult for machinists to establish legitimacy in the eyes of production workers. The maintenance-first strategy may have had unintended consequences, giving production workers a false sense of security.

Moreover, the machinist-first strategy may have backfired, as the internal power dynamics of the plant preserved a stark division between machinists and ordinary production workers. Layers of privilege generated ugly forms of resentment as solidarity struggled to cut across the skill divide. The machinists faced some unique issues that may have further distanced them from their production counterparts. One worker articulated this feeling of resentment tinged with jealousy:

> The way the line workers see it is their [the machinists'] job isn't as physical. It's obviously more educated. It's very hard to work on the line there. You're sweating, and you're breaking your body and having to have surgery just to keep working. And maintenance workers are standing next to the line drinking coffee. It can be difficult to see that and still understand why they get paid so much.[52]

Collusion and Labor-Management Partnership

The card-check procedure is a voluntary process that requires the cooperation of management. However, the UAW's push to achieve recognition through card check was almost immediately attacked by critics as a "secret deal" that would undermine workers' right to a free and fair election.[53] Union opponents reading the tea leaves at the time believed the UAW's fate was all but assured: VW sought to usher in the union under cover of the works council.

Yet early comments by UAW leadership immediately indicated that the organization was not gearing up for a traditional union drive. Even before organizers hit the pavement, the UAW was already handicapping its prospects. For his part, King sought to portray VW as a strategic ally and effusively praised the company's labor policies:

> They're the ones that most walk the talk. All the companies say they respect workers' rights to bargain, but then they let their American management run wild and violate workers' rights. VW doesn't do that.[54]

The UAW sought to frame its campaign for unionization as a path toward maximizing shareholder value, consistent with the union's drift away from its combative past. King justified this conciliatory approach by explaining that a UAW-VW partnership would "contribute to the company's success" and "benefit shareholders."[55] When confronted with what some saw as deferential language, he doubled down:

> Our philosophy is we want to work in partnership with companies to succeed. With every company that we work with, we're concerned about competitiveness.[56]

To King, VW's history of cooperative relations with German workers suggested it could be a different kind of company that would respond to a different type of organizing.

Once a union announces its intention to file, election drives typically ramp up and push vigorously for a favorable outcome. But even after filing, the UAW stood its course and maintained its "low-key, low-budget" approach, still banking on an easy win and sensing no real opposition. Moreover, the UAW still hoped it might persuade VW to deliver on card check, so it trod cautiously lest it antagonize the decision-makers. At this point, many workers felt that the UAW was colluding with VW. This perception was rooted in a grain of truth since, as all parties have acknowledged, the union and the company were holding private discussions. In the eyes of some workers, the "secret" talks amounted to something more sinister: an effort to build a de facto company union with no authentic rank-and-file participation. Most troubling, though there was extensive coordination behind the scenes, rank-and-file workers always seemed to be excluded from these arrangements.

The UAW might reasonably object that its backroom strategy was only one front in a battle with multiple theaters of engagement. However, in a zero-sum game, deploying business agents to cut deals with VW necessarily implied less attention to members. As one recalled:

> [The UAW] chose to court the company instead of the workers. It actually hurt us worse than helping us, them dealing with the company. When they started trying to deal with the company . . . they lost confidence in the union because the union and the company were already dealing together. And to the exclusion of the workers.[57]

The budding partnership between the UAW and the works council may have set the stage for an attenuated campaign in which both sides believed a union victory was all but preordained. As one described:

> The UAW, I believe, came in here thinking it was a done deal. Because they'd already had a stamp on it before it even started, they had the backing of the works council and IGM.[58]

Others claimed the UAW's collaborationist strategy was intended to lighten the load on staffers in a corner-cutting approach:

> [The organizing director] is dealing with the company to set up a framework so that the UAW has to do as little as possible to get their dues.[59]

Still others argued that the collaborationist strategy resulted in the demo-bilization of the membership:

> Regardless of what deal they may have with the company, they have conditioned us to have our union delivered on a silver platter by them and the works council.[60]

Much to the anti-unionists' chagrin, the UAW's access to the plant only seemed to grow in the crucial final week leading to the vote. Though the opposition went practically unchallenged in its publicity campaign, it could not compete with the union on the shop floor. Management afforded the UAW unfettered access to employees but made no similar accommodation for the opposition. Indeed, this was one of the most unusual features of the drive, breaking sharply from convention. Though the opposition had accused the UAW and VW of collusion from the beginning, the final pre-election week seemed to provide incontrovertible proof that something nefarious was afoot. While the company pointedly withheld any support for the UAW *as such*, critics on the right once again saw VW's mere cooperation with the UAW as a sign of nefarious behavior.

Union Oligarchy and Suppression of Dissent

Perhaps the most damning charge leveled against the union was the notion that it deliberately sowed divisiveness within the organizing committee. In the face of an intense, aggressive opposition campaign of ambiguous origins and unknown proportions, under the supervision of Regional Director Casteel, the UAW ran a staff-driven campaign and strove to maintain a facade of unity at the expense of cultivating local leadership, tolerating reasonable dissent, or fostering internal democracy.

Bitterness ransacked the campaign in the weeks leading to the vote, and the UAW seemed to cannibalize its organizing committee by diverting energy toward a hunt for the traitor within. Even as it locked horns with its corporate nemesis, the UAW was fighting an internal war against dissent. Two significant organizers left during the drive, and another two quit under duress. Finger-pointing had already begun within the organizing committee, but the UAW did nothing to stifle and much to encourage these accusations. Worker-organizer attrition adversely affected the drive and its aftermath. By firewalling the opposition, the UAW sent a message that the anti-union activists were not worth engaging.

If the UAW's internal operations were chaotic, the public face it presented to the world was disciplined and orderly. Staff insisted on approving all contact with the media, particularly as the stakes were raised during the late stages of the drive. In public appearances, workers projected confidence and unity, staying on message and relying heavily on talking points. This concealed from the outside world the internal discord that had begun to plague the campaign.

All union drives require a degree of coordination by professionally trained staffers. But organizing staff actively worked to suppress any independent initiative, far exceeding what might typically be expected from conscientious organizers. For example, two members suffered retaliation for attending a *Labor Notes* conference in Detroit for reasons related to the organization's history as a thorn in the side of the UAW. In addition to facing accusations of dishonesty, staffers were blasted for their megalomaniac tendencies. Workers felt that staffers demanded absolute deference from even the most independent-minded rank and filers.

Learning

In addition to poor access to knowledge and low motivation, the 2014 election suffered from an inability to adapt to changing conditions. As the earth shifted beneath its feet, the UAW seemed unable to adjust its plans accordingly. As the February election approached, members found themselves stuck with an outdated road map that no longer corresponded to conditions on the ground and was increasingly irrelevant to the challenges they faced. The union did not know what to expect, was not prepared to confront emergent threats, and retreated to a stunned deer-in-the-headlights posture even as it became increasingly outmatched. In short, lacking a viable organizational structure, the UAW was left with little recourse when its closed-door sessions failed to bear fruit.

Beyond hurting its election chances, the UAW's dereliction of duty exposed more endemic problems that stemmed from a faltering institutional memory. Rather than drawing hard lessons from its own experience organizing southern auto transplants, the UAW tried to reinvent its tactics on the fly, believing—as did many—that a "unique" set of circumstances demanded a wholly new approach. Had the UAW adopted tried-but-true organizing best practices or even adhered to its standard campaign playbook, it might have performed better. Instead, it ran an anemic campaign that relied on the supposed good graces of what it imagined to be an unconventional firm while ensuring the active demobilization of a workforce it assumed to be incapable of authentic self-organization. Ironically, though the UAW genuinely believed it was breaking new ground, its efforts to realign its strategy proved ineffective and counterproductive. Once VW peeled back its shiny veneer and the opposition mobilized, the union was left floundering—thoroughly delegitimized and without recourse. Indeed, when confronted with unexpected obstacles, the UAW often made precisely the wrong choices, magnifying rather than minimizing the obstacles themselves.

As time passed, rather than build momentum through active learning, the campaign stagnated. Union supporters felt—not without some justification—that they were in a state of purgatory through much of 2013. The campaign to collect cards had ended in the spring, and UAW staff led union supporters to believe either that they had a majority or that the cards were in New York awaiting an official count. (The latter claim was almost certainly false.) During the subsequent nine months, communication broke down, and workers were left to speculate whether VW had reneged on its commitment to honoring card check (as ultimately proved to be the case). Meanwhile, this period was characterized by a growing sense of impending doom. Apart from realizing that VW might not honor its end of the deal, the opposition, which had been a quiet minority in early 2013, became an aggressive anti-union campaign with national backing that rivaled the UAW in organizational effectiveness by 2014.

Activists anticipated that the UAW might assume a more aggressive position after its defeat in 2014. But old habits die hard, and even as the machinists worked to get organized, the union continued to pull its punches. Reflecting on the union's inability to change course, one worker said:

> After the first vote, they definitely should have changed strategy because it was apparent that their nonantagonization strategy had hurt. After the initial vote, I think they should have taken the glove off, so to speak, and come out a little harder, but they didn't. I think part of the problem was that they tried to find internal reasons to blame for the failure rather than, you know, faulting their own strategy.[61]

One worker went a step further and posed a counterfactual scenario, suggesting that a victory by an antagonistic UAW would have been bittersweet since it would have increased tension and animosity in the plant, creating a conflict-ridden and unpleasant work environment:

> Had it worked, it would have been a good strategy because then we would have been on friendly terms, people still would have been speaking to each other in the plant, and we would have been able to move forward more quickly. But it didn't work. And from that point forward, from the first vote forward, that strategy of "we're going to take VW at face value, and they told us they weren't going to do this" was quite possibly the dumbest thing they could have done.[62]

Workers often felt that the UAW's collaborationist approach led it to ignore VW's transgressions. As one wrote about a policy in which VW disallowed unscripted questions at team meetings:

> If this doesn't prove to the UAW that the Germans are two-faced sandbaggers, then nothing will. . . . When will the UAW wake up and realize that they are being led around by the short hairs?[63]

The union's passive strategy may have been linked to its initial expectation that it would easily unionize the plant. Since the UAW expected to emerge victorious, it did not want to burn bridges.

But others felt that the campaign's unassertive character was debilitating. If it initially appeared promising, the union's passive strategy did not place it on solid footing once VW reverted to its anti-union stance. One said ruefully, "When they showed their true colors, our hands were tied."[64]

Finally, and most importantly, the UAW's campaign suffered from path dependency: its inability, and at times refusal, to make necessary corrections when its preordained strategy sent it veering off course. Organizations tend to lock in certain behavior patterns and economic specialization. While lock-in can increase returns and competitive advantage, it can also become a barrier to change.[65] An organization capable of heuristic processes might have reassessed its strategy as circumstances dictated, but the UAW demonstrated no such resiliency. If a low-key approach seemed to make sense early on, that was no longer the case once the UAW confronted a bevy of unforeseen obstacles: anti-union campaigning by VW managers, organized opposition by politicians and other influential local players, aggressive and at times unlawful anti-union conduct by supervisors, and a company that drifted considerably from its early pro–works council (and tacitly pro-union) stance. Many workers expected that the gloves would come off once VW moved away from its early commitment to partnership. But in the weeks that followed the initial vote, as the consequences of the UAW's strategy became more glaring and increasingly impossible to ignore, the UAW surprised even some of its most ardent supporters by continuing to cooperate with VW as if it were still a reliable ally. The union, therefore, bears some responsibility for its own defeat in that it never laid the groundwork for a full-fledged organizing campaign even as the prospect of a coequal union-company alliance dimmed.

A counterfactual scenario is not difficult to imagine. From the standpoint of labor, the paradox of the corporate chameleon may even represent an opportunity of sorts. Entities that are unstable, erratic, or caught between two extremes are often defenseless against their foes. The gap between VW's stated mission and its practices may have presented vulnerabilities that savvy unions could exploit. The UAW's inability to exploit the fissure between Wolfsburg and Chattanooga was another unforced error. Too often, the UAW was duped by VW's chameleonic camouflage. The record shows that it should have known better.

Innovation

On our final measure of strategic capacity—innovation—the UAW earns slightly higher marks. Though the UAW in its heyday had corporatist aspirations, the last seventy-five years of case law have cemented individual rights as the overarching logic of the labor movement. Against this backdrop, the notion that a union might somehow call itself into being via a works council defies expectations, breaks with convention, and essentially has no precedent: it is the very definition of an innovative practice. By all accounts, this was an unusual arrangement that had never been attempted.

But innovation cannot generate strategic capacity on its own. Innovation must be integrated with the reflexive dimension of heuristic practices, recursively assimilating the past to move forward instead of desperately flailing in the abyss. Backed by knowledge, motivation, and learning, innovation provides creative solutions to carefully considered problems. Though staffers pitched the works council as a novel idea, it was completely divorced from the UAW's past experiences, the parameters of U.S. labor law, or, for that matter, its on-and-off alliance with IGM. Separated from its twinned heuristic process (learning), innovation cannot stand alone.

Thus, the UAW's first and only serious attempt to break from its path dependency perpetuated the very problems that had mired the organization in the first place. By courting the German works council while neglecting its prospective members in Tennessee, the UAW short-circuited widely accepted best practices and drew unwanted attention to democratic failures that were already contributing to its cycle of defeat. Thus, though 2014 had the appearance of strategic innovation, it backfired spectacularly. To the extent it occurred, innovation remained superficial and partial.

Conclusion

Measured against Ganz's four criteria, the UAW's organizing drive failed at multiple levels in 2014, and its defeat is no surprise. The primary explanation for the UAW's defeat is its failure to adequately prepare for known threats and its subsequent failure to readjust its strategy as circumstances dictated. Facing off against what at first appeared to be a "friendly" firm, the UAW eviscerated its unionization drive, coddling the employer rather than organizing members. Union supporters were eager for direction, starved for leadership, but ultimately abandoned by staffers who seemed more interested in currying favor with German executives than in developing worker organizers in Chattanooga. The union's dominant strategic approach was mismatched with the reality it confronted on the ground. As we have seen, the UAW took a soft-pedal approach to campaigning, replacing shop-floor mobilization with behind-the-scenes negotiations and voluntarily abstaining from nearly all of the persuasive strategies that typically constitute an organizing drive. Rather than make a convincing case on its own merits, the UAW presented itself as little more than a waystation en route to the ultimate goal of establishing a works council. Meanwhile, an aloof organizing staff and constant communication problems compromised any efforts to build internal leadership and engage the rank and file. The UAW was content, it seemed, to let the boss organize its members. Even more seriously, the union neglected to meaningfully change its course even when its tactical inadequacy became painfully obvious.

As PRA would have it, organizing drives derive their power and authority from a combination of associational power (drawn from strong shop-floor organization or community connections) and institutional power (based on connections to elites and levers of political influence or legal rights), structural power (derived from workers' strategic positioning within the broader economic system), and societal power (based on accumulated moral authority). Though these forms of power rarely exist in equal measure, a strategy overly reliant on one form of power over the others may reveal basic weaknesses as campaigns struggle to adjust to changing conditions on the ground. This was the trap that undercut the UAW's efforts. Lacking much structural or societal power and unable to build associational power, it gambled on institutional alliances—an anticipated agreement with VW, a partnership with the works council, and expected support from IGM. When these institutional connections collapsed or failed to fulfill their promise, the UAW was left without a backup plan. This reveals the dangers of treating institutional power as a replacement for associational power and serves as a cautionary tale for union leaders who may seek shortcuts to organizing. Even "friendly" employers cannot be trusted to respect basic union security, so unions that neglect to build an independent (associational) power base do so at their peril.[66] Whatever the merits of institutional power, it cannot substitute for member engagement and shop-floor mobilization.

4

Solidarity's Enclosures

The Dilemmas of Transnationalism

This chapter pulls back the analytical lens to reexamine the 2014 election from the standpoint of labor transnationalism, shifting our attention from the minutiae of organizing in Chattanooga to the union's connections, affiliations, dependencies, and alliances across space. Just as the UAW's company-friendly, corporatist orientation weakened its strategic capacity on a local scale, similar problems hindered the union's efforts to extend its influence and form partnerships across national lines. Under the entrenched leadership, institutional might had come to substitute for rank-and-file mobilization, directives were issued by administrative fiat, and deals were struck in corporate boardrooms, with workers often reduced to passive observers. Therefore, it should be no surprise that when operating on a transnational scale, the UAW sought relief through bureaucratic fixes that mirrored its own internal organizational tendencies. Referencing the company's decade-old pledge to respect the right to organize, the UAW appealed to VW's sense of moral virtue—even though such sentiments had proved empty in the past. Desperate for a toehold in Germany, the UAW held high-level talks with the works council, even though workers were shut out of these discussions and failed to see their value. Ultimately, any attempt to build transnational solidarity was hampered by organizational dysfunction and blinkered leadership that saw healthy deliberation as a threat to its centralized control.

The UAW's transnational project focused squarely on VW while inhibiting solidarity at a sectoral level and empowering the GWC, thereby shutting out the German trade union IGM. Its leaders exercised a form of "affiliatory control," or what I term *solidarity arbitrage*, reflecting and often perpetuating their antidemocratic and oligopolistic governance style. If labor transnationalism has the ostensible goal of growing networks and building connections, it also has the power to exclude relevant actors through its omissions and enclosures. The process through which otherwise well-meaning actors enter into rival solidaristic alliances gives rise to competitive pressures, subjecting employee organizations to horse trading as different groups jockey for competitive advantage on a varied terrain. Against more optimistic views of network effects, solidarity arbitrage exposes the dark side of labor transnationalism.

This chapter begins by tracing the recent history of social governance, pointing to the ineffectiveness of established regulatory tools while highlighting the potential of innovative new regulatory strategies. Drawing on transnational labor scholarship, it then proceeds to a discussion of global framework agreements (GFAs), emphasizing not their formal aspects but their indirect, second-order effects (i.e., the relationships, alliances, and connections they enable).[1] The extrajudicial nature of GFAs encourages the formation of transnational union networks while ensuring that the character and composition of these alliances remain contingent and subject to negotiation. With this theoretical background as its point of departure, the remainder of the chapter shows how a global union federation (GUF), a works council, and a company triangulated their strategy around a GFA against the backdrop of a contentious union election.

The Challenge of Labor Transnationalism

Considerable research has shown that worker competition drives down standards and undercuts solidarity, but the disunity and insularity of the world labor movement have long hampered efforts to regulate global production through labor transnationalism. For many years, international law was imagined as the logical mechanism for labor transnationalism. However, international law has proved weak to the point of uselessness, for it is routinely superseded by local and national law. Deeply held traditions of national embeddedness, organizational inertia, and short-termism—along with practical considerations of concerted action across vast distances—have often reduced "labor transnationalism" to an empty slogan.[2] Whipsawing (in which management plays plants against each other to extract concessions, typically by lowering standards at one plant to increase pressure on another) has further undermined transnational unionism.[3] Given that labor transnationalism almost definitionally requires high levels of cooperation between aggrieved stakeholders, divisiveness and animosity between labor organizations remain major impediments to achieving meaningful global regulation.

Considering the elusiveness of effective regulation, it is understandable that scholars would be keen to highlight and celebrate tactical victories. However, new strategic alignments create unplanned risks and vulnerabilities, which, left unchecked, may hinder labor's long-term advancement.

Global Framework Agreements

In the absence of a global legal framework for negotiating, implementing, and enforcing labor standards, it is unclear precisely what form meaningful labor transnationalism might take. Thus, existing examples of effective solidarity tend to be either one-off efforts that succeeded against the odds or highly irregular configurations resulting as much from luck as from strategic foresight. Since the legal architecture that might allow for the creation of permanent multinational labor institutions essentially does not exist, the question of organization remains a salient issue. Given the aforementioned challenges, it is often taken as axiomatic that social governance remains in its conceptual infancy,[4] is only beginning to emerge,[5] has yet to gain its footing,[6] or has a sporadic and piecemeal character.[7]

Yet while such a fatalistic conception of social governance may have roughly approximated the state of affairs in the late twentieth century, it is increasingly outdated.[8] Labor transnationalism is no longer a mere aspirational ideal. Unions and their allies have proved capable of successfully countering the tendency toward whipsawing through a revanchist program of transnational solidarity (variously referred to as "political entrepreneurship,"[9] the "boomerang effect,"[10] or "reverse whipsawing"[11]). Labor has also sought to codify social governance standards across divergent regulatory environments through innovative policy tools known as GFAs.[12] Since their inception, these agreements have garnered significant attention from labor academics, offering a potential retort to capital's "spatial fix" by way of an administrative fix.[13] GFAs aim to give labor an active role in global production networks[14] and can aid in the institutionalization of social dialogue[15] and the local-level implementation of private labor standards.[16]

Few weapons in labor's arsenal have generated more enthusiasm and interest than GFAs. It is perhaps no surprise that the American labor movement has been eager to embrace GFAs as a bureaucratic fix to the problem of transnationalism. GFAs offer a solution to the territorial dimension of labor's scalar dilemma, representing the ultimate expansion of labor's national reach. GFAs also appeal partly because of their relative simplicity: they extend a basic set of standards across a far-flung corporate empire, replacing the vagaries of local customs and regulatory regimes with a universal and eminently desirable objective. But like other aborted attempts at internationalism, GUFs can be ill suited to a highly differentiated world in which regional specificity is

naturalized through habitus and buried beneath layers of ideological justification. Given the backdrop of labor's precipitous decline, GFAs seem to offer
an alternative that implies a break with the past and, by extension, a reversal
of labor's (mis)fortunes.

Limitations

Despite their name, GFAs are not necessarily global, routinely fail to establish a coherent framework, and rarely signify total agreement. Instead, they
must be understood as highly contingent documents that firms read as mere
suggestions. A GFA's capaciousness is also its chief vulnerability. Their guidelines are notoriously difficult to enforce and rely primarily on a company's
goodwill and concern for its reputation. An entire cottage industry has persuasively argued that, with rare exceptions, GFAs have little material impact.
Though GFAs have a twenty-year history and are now approaching their late
adolescence as regulatory mechanisms, serious questions remain about their
effectiveness. They are virtually meaningless at nonunionized firms, and
some conclude that unionization is a prerequisite for successful implementation and enforcement of compliance.[17] The voluntaristic character of GFAs
is among their chief liabilities: without a firm's cooperation, the GFA becomes
moot in practical terms.

Core provisions enshrined through GFAs are routinely ignored with impunity. Nonunion subsidiaries often fail to benefit from the paper commitments made by their parent companies, especially those concerning freedom
of association and the right to collective bargaining.[18] Adding to the problem,
international bodies like the International Metalworkers' Federation (IMF)
lack enforcement capabilities and remain wholly dependent on the voluntary initiatives of their national affiliates. Therefore, while GFAs have proved
effective in tightly linked logistical industries like maritime shipping, where
the taut production chains and deeply integrated choke points promote cooperation and codependence among national unions,[19] they have been less
influential in manufacturing, even as mass industry rapidly shifts to just-in-
time production.

From a juridical-legal perspective, GFAs are largely symbolic. As Christina Niforou puts it, "GFAs stipulate global commitments which in practice
are subject to local laws and institutions."[20] Further, despite optimistic pre-
dictions that GUFs might someday claim representation of employees in
company operations worldwide, this legal theory has not been tested and
seems to push the boundaries of jurisprudence. Though International Labor
Organization (ILO) conventions may appear to be a copy-and-paste set of universal standards that have already achieved broad consensus, in practice, they
are no match for local laws when the two come into conflict.[21] Indeed, in those

rare cases where local actors successfully persuade a TNC to exceed the minimum standards mandated by the host country, it is typically internal pressure and the threat of disruption, not the mere existence of the GFA itself, that force the company's hand. While the GFA may provide local actors with ideological cover or normative justification, its inherent value is unclear. Put differently, a literalist interpretation of GFAs is prone to misattribution of causality: even when GFAs are linked to pro-union outcomes, the GFA may not be a causal mechanism but rather an effect of a preexisting union-friendly relationship.

Given the fact that enforcement is largely at the discretion of TNCs, some cynical scholars argue that GFAs are little more than window dressing, akin to corporate codes of conduct used by TNCs "to protect themselves from negative publicity, foster public relations with unions," and tap into "socially responsible niche markets."[22] Because of the weakness of GFAs, transnational union cooperation tends to be achieved through informal or loose networks rather than through formal, institutionalized arrangements.

However, it is beyond the scope of this chapter to rehearse well-trod arguments about GFAs as formal documents. Instead, we are more interested in how GFAs rearrange the social field, reposition key actors, and alter power relations. Unfortunately, much research on GFAs hedges toward legalistic formalism while remaining largely indifferent to practices of power that enable or restrict legalistic solutions.[23] Moreover, as critical legal theory would have it, policy is not a fix but a contested terrain of struggle in which both the phrasing of an agreement and its impact on the ground are subject to negotiation in real time.[24] Therefore, the discussion of GFAs that follows focuses less on their manifest content than on their capacity to inculcate organizational routines and (re)structure the social field.

GFAs tend to bear the imprint of their lead signatory: most often GWCs.[25] To draw a broad generalization, the proximity of works councils to management reduces their militancy and increases their cooperativeness compared to unions. Unlike unions, whose drift toward accommodationist politics was always contested and contingent, works councils exist to preserve industrial peace. Therefore, it might be no surprise that among the loudest defenders of GFAs are those who explicitly adopt an institutional framework.[26] Unions, in contrast, may be openly skeptical of GFAs.

Though labor internationalism aspires to exist in the ethereal nonplace of global capital, GFAs are corporation based and risk segmenting the labor force.[27] In Europe, where the tradition of industrial unionism is even more entrenched than in the United States, pattern bargaining is an accepted, normalized practice. GFAs represent a 180-degree retreat from this tradition, a return to the firm as the purveyor of rights and locus of struggle, even if in the name of globalism. GFAs may strengthen ties between headquarters and subsidiaries within a single company, but they do so at the expense of weakening the potential for sectoral (multifirm) bargaining. As Dimitris Stevis and Terry Boswell caution:

> A very real danger of strategies that focus on individual companies rather than taking a broader view is the fragmentation of unions across corporate or geographic lines. In the case of agreements, workers may end up caring more about the competitiveness of their company and less about union power.[28]

GFAs incentivize workers to identify their interests with the firm and see them as bound up in the viability of their employer's expansionary prospects.

While the headquarters performs certain organizational functions, it is not necessarily the command center of the firm, nor is it the locus of power from which control descends. Instead, it is one node of many in a decentered organization. Policy must be filtered through subsidiaries (e.g., VW Group of America) and plant-level management, so its impact may be diluted by the time it percolates through layers of bureaucracy and reaches the shopfloor. Moreover, subsidiaries are not signatories to GFAs; thus, their conduct depends on headquarters' ability to persuade them to comply.[29] GFAs assume close integration between headquarters and subsidiary units (or subcontractors), but modern TNCs are internally differentiated and heavily stratified. Even if organizational charts indicate that the headquarters bears ultimate responsibility for the actions of subordinate levels, the headquarters may not be properly equipped to oversee its underlings.

Following actor-network theory's challenge to organization studies, "the firm" is not necessarily a cohesive entity that can be expected to behave in a predictable way given certain prerequisites. Instead, the firm is a complex set of institutions and practices ranging from the signing of GFAs to the calisthenics instruction that opened workers' day at the Chattanooga factory. These practices are constitutive components of the overall arrangement that comprises the firm, occurring in places without necessarily being defined or delimited by those places.

GFAs are often heavily based on the culture and institutional or legal framework of the TNC's country of origin; this is reflected most clearly in the fact that home-country unions typically overshadow all others in terms of national union involvement. If GUFs have often struggled to attract meaningful participation from non-European members, this problem is revealed even more starkly with respect to GFAs.[30] In the American context, where soft-law regulations have not succeeded in reining in corporate power, unions may fail to see the added value of a GFA unless it is directly tied to a comprehensive cross-border campaign.

GFAs have come under attack for their allegedly antidemocratic char-acter.[31] For many critics, GFAs increase the distance between the shopfloor and corporate headquarters. Though persuading all affected parties to assent to a GFA can be contentious, these debates occur at considerable geographic and organizational remove from the shop floor. Worker input, to the extent it is tolerated at all, is mediated by high-level representatives (or, in the case of GUFs, second-order representatives) who have little day-to-day contact with the rank and file. This issue has knock-on consequences for effectiveness: guidance from above that is disconnected from reality on the ground and insensitive to local conditions may backfire when the rubber meets the road. Moreover, GFAs are often couched in the highly abstract language of "eth-ics," further evincing their irrelevance to grounded reality. Likewise, once a GFA is approved, the inevitable ensuing struggles over implementation, compliance, and oversight are detached (both physically and administra-tively) from the executive suites in which the GFA was formulated.

If anything, GFAs may contravene the kind of shop-floor militancy and deep person-to-person organizing that has been shown to deliver results in a U.S. context. GFAs are located "at the peak of a pyramidal structure sev-eral removes—and gatekeepers—away from any flesh-and-blood workers."[32] GFAs are structurally exclusionary tools that reduce rank-and-file workers to passive recipients of their leaders' beneficence.

U.S. labor relations are predicated on an adversarial model: the idea that optimal outcomes emerge from conflictual deliberations between self-inter-ested parties. GFAs, to the extent that they are conflictual at all, take place in the register of labor law, not on the terrain of shop-floor struggle.[33] What sepa-rates positive examples from negative ones is the presence of on-the-ground rank-and-file activism, not merely as a supplement to top-level negotiation but as the life-giving force that invigorates what is otherwise little more than a statement of intent. In the most generous telling, GFAs are a battleground be-tween unions and companies that creates space for contestation and opens an arena of struggle and possibility.

In summary, GFAs center the works council, the home country, and Eu-rope while decentering unions, workers, and the host country, all the while undermining democracy. Yet despite these strident critiques, there are rea-sons to view GFAs more favorably.

Strengths

Though their effectiveness as formal policy remains dubious,[34] GFAs have had perhaps their greatest impact via their indirect, knock-on effects. Even when they prove ineffectual, GFAs call into being an emergent arena that reinvigo-rates existing federated structures, encourages the forging of informal alliances between previously isolated actors, and promotes realignment among affected parties.[35] Thus, regardless of their value as regulatory tools, GFAs have altered the landscape for global labor by giving rise to a plane of contestation in which multiple actors enter into strategic partnerships and jockey for position across a variegated terrain, with lasting consequences for the project of labor transnationalism. In shifting our focus away from its manifest content and toward its processual dimensions, we might evaluate GFAs more favorably.

Though their quasi-contractual language might suggest that GFAs established a new legal doctrine out of whole cloth, this is not the case. While many studies have examined and compared GFAs as texts, given their provisional character, ambiguous legal status, and inconsistent application, it may be more helpful to view GFAs as the living record of an emergent process that is never fully codified. Indeed, GFAs are provisional arrangements that result partly from the incongruity between actors and the absence of established relationships between relevant parties. This process may generate documents that carry the air of finality, but it is, in fact, subject to constant negotiation and renegotiation. The constellation of actors co-constituting a GFA is organizationally complex and subject to the challenges of multilevel governance: worker representatives may coexist at the global, pan-national, national, company, and local levels; employee representation may oscillate between trade unions and works councils. For these reasons, GFAs may be something of a Trojan horse that enables a range of interactions, none of which have been explicitly provisioned by the GFA, but few of which would be likely to happen in the GFA's absence.[36]

Markus Helfen and Jörg Sydow further advance our understanding of GFAs by pointing to their unsettled, contingent, and contested character.[37] Institutional actors are not ironbound by their institutional context; they have some freedom to determine whether they will comply with expectations or break from tradition. Thus, while context can be understood as a determinant of outcomes, it must not be seen as determinative. Despite gaining popularity, GFAs are still untested doctrine, and their reach and purview remain largely uncharted. Thus, it is unsurprising that negotiations over GFAs may be time consuming, with false starts and failures, particularly when they broach new fields and arenas (such as works councils in the United States). The relationships, connections, and interactions that enable the negotiation of labor-friendly social contracts are as important as their outcomes. Negotiation work may generate proto-institutions—that is, arrangements that remain provisional and unsettled but emerge from a dynamic process subject to renegotiation. Consistent with this call to consider the processual dimensions of GFAs, Michael Fichter and Jamie McCallum suggest that the strongest GFAs are contested through a "conflict partnership," which they present as an alternative to "social partnership." In their view, a GFA forged through both "battle and dialogue" will have more staying power and prove more enforceable. Put differently, the institutional power a GFA represents is indebted to—and indeed dependent on—the associational power from which it derives.[38] This was not the case with VW's GFA, which instead exemplifies what Richard Hyman calls "the fatal attraction of the elitist embrace," or the tendency for labor-friendly firms to demobilize unionists and undermine their power by baiting them with the (false) promise of cooperative relations.[39] As Fichter and McCallum write, "Enlightened management, one that is generally supportive of securing the GFA and indicative of a social partnership approach, might ultimately be a large factor in ensuring that it is never implemented."[40] Though VW assented to a GFA fairly easily, convincing the firm to abide by the agreement was a very different matter. Again, the divorce of the global from its local dimension—on either side of the Atlantic—made VW's GFA a futile exercise in labor statesmanship.

Synthesis

Evaluating competing claims about GFAs will require moving beyond a strict legalistic framework. Similarly, following a systematic review, Marc-Antonin Hennebert, Isabelle Roberge-Maltais, and Urwana Coiquaud conclude that "the main impact of GFAs is the stimulation and development of social dialogue," leading to a virtuous cycle that encourages more widespread conversations, eventually resulting in salutary outcomes.[41] On the other hand, such second-order effects are difficult to gauge or measure objectively. Let us grant for a moment that GFAs really can become "levers of influence and opportunities" that "provide reference points and spaces to develop international links at critical moments," as their proponents maintain.[42] If true, these claims can only be evaluated by examining first-order impacts. "Opportunities" become apparent when they are seized, and "spaces to develop international links" are not self-evident until the alliance is consummated. The conscientious social scientist is then faced with the unenviable task of backfilling a counterfactual scenario: how might outcomes have differed had the "opportunities" and "spaces" not been created? Proving that GFAs contribute to positive outcomes (albeit indirectly) requires an evidentiary trail connecting cause to effect. Up to this point, though plenty of anecdotal information suggests GFAs may be copresent with solidaristic alliances, such circumstantial evidence stops short of indicating a causal relationship. Qualitative social science need not be held to the strict scientific standard of experimental design, but neither should it speculate about "opportunities" and "space" that can only be revealed through their second-order aftereffects. Additional empirical research is required to determine how GFAs interact with other strategies and tactics in a dynamic field.

Works Councils and Labor Transnationalism

Traditionally, transnational labor pivoted on the trade union, which became the base unit on which a cross-border alliance might be constructed. But under GFAs, GWCs are centered as lead actors while pushing unions to the margins. This section critically evaluates the GWC as a transnational agent.

Despite achieving near-mythological status in the eyes of the American Left, GWCs have a mixed record extending their influence beyond Europe. Unlike Toyotaism, which was initially adopted when *kaizen* became a buzzword in the United States and later in management circles worldwide, codetermination and other aspects of the German system never gained traction outside Europe. VW exported its works council abroad as a vestigial remnant of the company's origin, preserved as a cultural signpost for a kind of aspirational "Germanness." For example, despite a local culture of militant unionism, the works council has been absent in Brazil, where it could not stop a deeply unpopular plan to integrate suppliers into core operations.[43] Even more glaringly, on several occasions, the VW GWC has actively undercut the possibility of solidarity between IGM and diasporic national unions. These are not scenes of betrayal or treachery, for IGM and the works council continue to present publicly as allies. However, when VW has faced shop-level conflict abroad, it has consistently been the works council, not IGM, that colludes with the company to tamp down industrial strife. For example, the GWC fended off attempts by a breakaway union in South Africa to mount an independent challenge to the National Union of Metalworkers of South Africa (NUMSA). After a dissident caucus within the NUMSA was denied ballot access, supporters organized a spontaneous wildcat strike, resulting in mass dismissals. The NUMSA refused to fight for the reinstatement of fired workers, with the express support of the GWC. In contrast, the GWC maintained that the breakaway faction consisted of troublemakers who "[did] not deserve solidarity."[44]

Structurally, the GWC is, at best, a semi-independent entity partially integrated into VW's management structure under the German codetermination regime. As critics note, GWCs are guilty of "collaborating in lowering wages and working standards for the periphery of the workforce . . . with the aim of defending the core workforce."[45] Moreover, like GFAs, GWCs are bound by the strictures of their firm-centered basis. Even in Germany, GWCs are sometimes seen as egotistically motivated, with less interest in improving conditions throughout the industry than protecting their own members. Their self-centeredness can inhibit solidarity at the sectoral level.

There is widespread disagreement over the place of works councils in German society. Some view them as a conciliatory engine of democracy, while others see them as a weapon of class conflict.[46] According to Ulrich Jurgens,

GWCs are tools for pacifying industrial unrest by granting employees a stake in their employer's success.[47] Indeed, the VW works council has historically been a critical pillar of stability within the firm, counteracting union militancy. Stephen Silvia writes, "Management has relied on the integration of the works councils into the firm's decision-making structure as a means to secure and retain employee buy-in, particularly for contentious decisions, such as layoffs."[48] There is widespread recognition among business experts that Germany's codetermination system has contributed immeasurably to the success of its major firms.

Yet in Europe, some scholars have suggested that works councils are approaching obsolescence, having long ago lost legitimacy.[49] These authors claim that unions often struggle to explain to members why a transnational identity is necessary. Workers instead prefer a plant-level or employer-level identity, which may be more salient. Despite their apparent popularity as judged by their membership rolls, works councils have lost institutional legitimacy over time and may no longer be suited to the challenges they face. This is not to sound the death knell for works councils, which remain among the most powerful and enduring institutions of labor solidarity. Indeed, any new formation will immediately run up against the former hierarchical order, which remains intact, supported by a network of institutions and actors capable of arguing in favor of the interests they serve.

The UAW and Labor Transnationalism

The American labor movement was slow to embrace transnationalism, and the UAW is no exception. In the 1980s, still reeling from the sting of deindustrialization and seduced by the revanchist chauvinism of Ronald Reagan's (original) MAGA posturing, the UAW embraced an uninhibited nationalism that slid easily into xenophobia. Famously, the UAW staged a form of primitive street theater that involved smashing Japanese-made vehicles with sledgehammers, culminating in the senseless murder of a Japanese American immigrant. If these events dominated media accounts, the UAW also fell prey to a subtler form of nationalism, less virulent in its style but just as protectionist in its logic, in which the UAW couched its opposition to NAFTA as a paternalistic defense of Mexican workers' rights. The UAW, in its condescending savior mode,[50] may have been less ugly, but it was no less problematic. In short, labor's resistance to NAFTA—in both its modalities—was characterized by poorly concealed xenophobia, an explicit privileging of the interests of American workers, and the liberal use of racist and nationalist tropes.

Additionally, any potential collaboration between the UAW and the GWC was hampered from the start by the history of the two organizations. While the GWC has aggressively pursued international alliances, the UAW has engaged in internationalism only sporadically, dutifully attending its GUF meetings but rarely forging or participating in durable relationships with overseas partners.[51] When offered the opportunity to engage in cross-border campaigns, the UAW has not risen to the challenge. Unions that pursue international strategies often pursue firms headquartered in their home country most aggressively, but the UAW falls short even on this count. For example, when the Brazilian union CUT (Central Única dos Trabalhadores) faced layoffs at a Brazil-based Ford plant, the UAW refused assistance, leaving CUT to negotiate independently with the automaker in Detroit. In a different context, the UAW declined to participate in a solidarity campaign to support workers facing human rights abuses in Venezuela despite pressure from the global community to intervene.[52] These examples stand in stark contrast to CUT's positive experience with IGM, which, with assistance from the GWC, helped Brazilians avoid job losses under similar circumstances at a VW plant.

Moreover, even the most cursory examination of the UAW's fitful efforts to build transnational alliances reveals a glaring failure to engage the local dimension, without which global initiatives are rendered ineffective. Ultimately, the UAW's "international" strategy—to the extent that there was one—could not stand alone.[53] As Michael Fichter argues, the UAW sought to substitute institutional power for its lack of associational power from the start.[54] This strategy was undermined by VW's ability to "rescale" the conflict to a different institutional setting—that of the South—where the GWC's institutional power was muted and the UAW's associational power held little sway.[55] Lacking the necessary pillars of an international strategy, the UAW could not turn its partnership with the GWC into an advantage on the ground.

Global Framework Agreements at VW

VW has long been a standard-bearer because it is the largest and most acclaimed representative of the "German" model, enjoying and cultivating a reputation for union tolerance compared to industry peers.[56] Indeed, VW is often treated as something of a synecdoche for the German labor relations system writ large, given its outsize influence. With its supposedly enlightened management practices, it is second perhaps only to Volvo in terms of its corporate social responsibility reputation. In this sense, VW functioned as a social laboratory for testing GFAs. VW's 2002 GFA broke new ground in its scope and comprehensiveness, while its 2009 side pact was the first to specifically recognize the rights of workers employed by subsidiaries and contractors. The IMF was also an early adopter of GFAs, signing fifteen by 2007 (more than any other GUF), the majority by German-headquartered firms. (As the IMF's largest and most influential affiliate, IGM was the driving impetus behind most of these agreements.) Because of this shared history, both the IMF and VW have established themselves as pioneers and standard-bearers in the burgeoning field of global labor agreements. This is a mixed blessing: when VW and IMF agreements fall short, suffer from uneven implementation, or fail to have their expected impact, the events are often interpreted as an indictment of GFAs tout court.

Despite the history of collaboration previously outlined, negotiations over VW's GFA circumvented both the IMF and IGM, instead positioning the company-centric works council as the intermediary. With this precedent established, the GWC quickly moved to present itself as the face of labor internationalism at VW. Though the UAW had a preexisting relationship with IGM, it shifted its focus to the works council, denying itself a key source of leverage. In fact, IGM was openly skeptical of the GFA, writing, "It is far better to have no [GFA] than a weak one."[57] UAW representatives traveled to Germany for talks with the works council and (separately) with VW's HR department. However, the HR talks were conditioned on the exclusion of IGM from these early conversations—a stipulation the UAW accepted to avoid alienating the company. Setting a pattern that would have consequences at VW and beyond, the company's strategic exclusion of IGM was problematic, as IGM was more relevant than the works council. Additionally, the GWC had more internal skill-based hierarchy than IGM, configured as it was on a craft basis and lacking IGM's formally democratic "industrial" character. Moreover, in public statements, the works council head explicitly embraced union substitution, presenting the works council as an alternative to and replacement for collective bargaining, even though U.S. labor law prohibits such an arrangement.[58]

Having emerged as the public face of worker internationalism, the GWC took the lead role in its push for employee representation in Chattanooga. Workers came to observe these shifting alliances firsthand. When workers came into contact with their counterparts from abroad, those were much more likely to be GWC officials than IGM representatives. For example, a select group of workers from the Chattanooga plant were invited to a GWC meeting in Stuttgart, but there was no parallel invitation to attend IndustriALL Global Union meetings in Geneva. Top-level works council figures frequently appeared inside the plant (with the company's blessing); IGM did not request and was not granted similar access. Statements of support from German workers consistently bore the imprimatur of the works council; IGM was less consistently included. And, perhaps most seriously, the unionization campaign itself was advertised to workers as a waypoint en route to the ultimate objective of chartering an American works council.

Rather than organizing a conventional union drive, the UAW presented the establishment of a works council as its primary goal and the creation of a collective bargaining unit as a stepping stone that might lay the groundwork for that overriding objective. Shorn of its militant character and tethered to the works council, for which it was a mere interstitial placeholder, the UAW was reduced to a works council-in-waiting and could not persuasively advocate on its own behalf.

Unfortunately, works councils have no formal standing under U.S. labor law, and their legality remains fiercely contested.[59] American workers have no positive experience with works councils, and the bid to solicit worker input faintly echoes the UAW's decades-long battle against the team concept popularized by the Japanese. Whether or not the comparison has any merit is beside the point; the U.S. experience with participation schemes turned a generation of workers against any policies that might bring them closer to management's orbit. Lacking knowledge of the intricacies of German labor relations and unfamiliar with works councils, Chattanooga workers were surprised and often confused by the works council's entreaties.

In sum, GFAs adopt a synoptic view of the firm that mirrors the managerialist perspective. The company is organized hierarchically, so solidarity among units, subunits, subsidiaries, and suppliers must be premised on the company's command structure. Thus, under the direction of the GWC, grievances over the company's conduct in Chattanooga were redirected to German headquarters. Believing that only German management could rescue them from American supervisors, unionists in Chattanooga built a strategy that depended on the headquarters' willingness to intervene.

Impact

As our case study makes patently obvious, GFAs are not, in themselves, a magic bullet that will clear the path for organizing. This reality is at odds with perception. But GFA defenders continue to exaggerate their capabilities, insisting against the evidence that they might pave the way for representation where none exists. Ironically, this is the task to which they are most poorly suited. When it comes to guaranteeing the right to collective association, GFAs are simply not an effective tool. Indeed, previous research has shown that GFAs are most effective in unionized settings.[60] GFAs can provide existing representatives with leverage,[61] but have a poor track record encouraging unionization in new contexts.[62] Thus, VW's GFA did not seem to meaningfully benefit the UAW. The company's paper commitment to freedom of association had no bearing on its conduct in Chattanooga.

Though many commentators were surprised by VW's apparent disregard for its GFA, existing research suggests this outcome should not have been entirely unexpected. Even under ideal circumstances, company compliance with GFAs can be inconsistent and faltering,[63] but certain factors reduce accountability. Among these, national-level anti-union policies,[64] the relative independence of local managers,[65] weak control by the head office over subsidiaries,[66] and a lack of coordination among various employee representative bodies are all major barriers to GFAs; each was present in Chattanooga. Conversely, factors known to facilitate the successful implementation of a GFA, such as an active membership and involvement from a GUF,[67] were largely absent, as the GWC displaced IGM and IndustriALL as the primary employee representative body on the European side.[68] Moreover, of the various provisions contained within GFAs, clauses covering the right to representation and voluntary recognition have proved the most troublesome; again, these were at issue in our case study. Finally, the applicability of a GFA to as yet nonunionized employees in the United States is unclear, even according to the ethereal legal standard of soft law. The literature predicts that without a GFA, European companies will not exceed the minimum labor standards tolerated by national-level regimes.[69] The expectation that American workers might be protected by a GFA to which they were not formally a party proved overly optimistic.

Defenders of GFAs often cite their indirect benefits, which result not from their express content but from the informal alliances they help forge. However, in the final analysis, GFAs must be evaluated on the basis of their measurable effects. In Chattanooga, the GFA succeeded in initiating a sustained dialogue, but this dialogue did not result in concrete practices on the ground. From the UAW's standpoint, the GFA was an objective failure, as it did not compel VW to abide by the minimum labor standards it outlined.

Evaluation

Meanwhile, rather than attending to affairs at home, the UAW banked on an outside savior, looking abroad to the GWC as a deus ex machina that might swoop in and save the day, rescuing workers from their plight. Transnational solidarity is no panacea, particularly in the absence of a solid organizational presence on the ground in the United States. The UAW overestimated the GWC's ability and willingness to force neutrality on VW. The UAW trusted its allies to pull the levers in Germany, banking on moral suasion and a GFA ("Declaration on Social Rights and Industrial Relationships at VW") while spending time and money flying workers back and forth to Wolfsburg for high-level meetings with GWC reps. The union had to have its house in order in the United States before it could position itself to benefit from outside help.

Ultimately, the UAW's half-baked attempt to forge a transnational alliance became a time-consuming and resource-depleting distraction from the real work of organizing. For all its appeal, efforts to forge a works council in Chattanooga bore a closer resemblance to an expository exercise in creative legal theory than a real strategy for mobilization. Moreover, works councils reached the apex of their power during the postwar era of relative stability and prosperity in the West—a moment that has long since passed. The constant dredging up of structures and organizations from the twentieth century shows the failure of the radical imagination and the unwillingness of labor and its allies to conceive of new formations appropriate to a time of global tumult and institutional decline.

Solidarity Arbitrage

Research on transnationalism has pointed to the importance of the spillover effects of union-inclusive initiatives that foster "trust and common group understandings," thereby facilitating and enabling additional initiatives of a similar nature.[70] Yet the logical corollary remains underexplored: union-exclusive initiatives may propel their own self-justificatory cycle, further reducing union involvement downstream. In cybernetic terms, if spillover is generally modeled as an accumulative process, its devolved variant may instead resemble a negative feedback loop in which the system achieves stability through decremental change.[71] This chapter has enriched our understanding of transnationalism by exposing its capacity to exclude, restrict, or otherwise impede union involvement.

The tenuous and ad hoc character of UAW's circa 2014 flirtation with transnationalism traces its origin to an earlier effort to implement a GFA in 2002. On paper, the IMF and the GWC appeared to be coequal signatories to each of VW's GFAs, but the GWC quickly established itself as the lead actor, with the IMF reduced to a subordinate junior partner.[72] Their relationship might be described as one of "differential inclusion": while the IMF had a seat at the table, in practice it deferred to the GWC on most GFA-related matters. This repositioning undercut the UAW's budding partnership with IGM while elevating the GWC. Though subtle, this pattern of differential inclusion eventually proved habit forming and had severe downstream implications. IGM and its corresponding GUF were effectively sidelined, while the GWC assumed center stage as the handpicked labor designee for VW's overseas concerns. The UAW was then forced into an arranged marriage with the GWC, a party with which it had no meaningful prior relationship. This alliance quickly proved unstable, and the UAW experienced a legitimation crisis, which contributed decisively to its ultimate defeat by an aggressive anti-union effort.

Arbitrage evokes the complexity of power relations in the global economy. Negotiation over social governance creates a political arena where institutional actors compete for influence, based less on the exclusion of weaker parties than on the varying distribution of power across the field. From capital's perspective, maintaining a dynamic arena where competitive pressures persist indefinitely is productive as it creates perverse incentives to play various actors against each other.

However, if GFAs open new opportunities for social governance, they also create new dilemmas for labor-side actors. Because GFAs are somewhat provisional and exist principally in the ill-defined realm of soft law, it is rarely obvious which actors are best suited to the tasks of negotiation, oversight, and enforcement. Potential signatories must be selected from a dense thicket of national and supranational organizational entities, possibly including GUFs, national union federations, national unions, firm-centric GWCs, and industry-based regional works councils, each containing internal hierarchies. Even assuming all parties act in good faith and prioritize shared interests, the selection process necessarily creates and maintains inequalities of power and access. Therefore, while GFAs provide a basis for solidarity, they also may exacerbate existing tensions by incentivizing competition over a scarce resource: influence.

VW was obliged to triage the demands of an American union (the UAW), a German union (IGM), a multisectoral GUF (in this case, the IMF, later IndustriALL), and a works council with global, European, and general (national) tiers. Though the various parties maintained a cooperative spirit and displayed no overt tension, this was a heterogeneous and internally differentiated assemblage of social actors with diverse motives, incentives, organizational structures, and institutional memories. Complicating matters further, these were not fully autonomous actors but nested entities with interlocking directorates and overlapping jurisdictions.

These dynamics reveal the two-faced character of spillover. While current scholarship on spillover across transnational agreements highlights the long-term benefits of "union-inclusive" initiatives,[73] this chapter adds a layer of nuance, proposing that even formal inclusion may obscure calibrated systems of access. If spillover is normally treated as an accelerationist process through which actors accumulate influence and become more deeply integrated over time, it can also result in the diminishment of capacities, a reduction of capabilities, and the severing of bonds and alliances. Though existing research on spillover has emphasized the productive and beneficial consequences of connectivity, spillover can also have unintended consequences, which may be harmful from labor's standpoint.

At the company's urging, the GWC, formerly a bit player, took the lead in coordinating the transnational alliance that had previously been the domain of IGM and its parent GUF. The GWC was an awkward negotiating partner because the primary issue (the right to organize) was beyond its purview and because it had no symmetrical counterpart in the United States, where GWCs are absent and have no legal standing. Though the strategic use of the (relatively compliant) GWC alongside a (relatively militant) union might allow for a good-cop, bad-cop dynamic that would gradually wear down the company's resolve, there is no evidence of an effort to coordinate in this fashion. In the final analysis, the existing alliance between the UAW and IGM did not translate easily or directly into a partnership with the GWC. The absence of prior contact between the UAW and the GWC impeded a meaningful alliance, and the opportunity for a deeper connection with the IGM was lost.

Conclusion

This chapter has shown that the UAW's strategy in the 2014 election fore-grounded the GWC and its GFA while sidelining the GUF and IGM. The partnership between the UAW and the GWC was premised on an instrumentalist notion of solidarity, in which the UAW was to be the benefactor of GWC beneficence. However, as long as the IMF was excluded, incongruity existed between the two bargaining teams, and the alliance suffered from scalar incompatibility. Specifically, VW's transnational character has no symmetrical equivalent since both the IGM and the UAW are national (or, in the case of IGM, predominantly national with some pan-European extensions). The real institutional equivalent to VW is the autoworkers' GUF, the IMF, a proto-institution not party to the bargaining. As a result, the UAW's organizing campaign took on the stylings and attributes of the GWC with which it was now paired, becoming (1) firm-centered, (2) accommodationist, (3) elitist, and (4) distinctly German. Had the UAW instead maintained and strengthened its partnership with the GUFs, it might have resulted in a vastly different campaign that was (1) multifirm or sectoral, (2) militant, (3) democratic, and (4) adapted to the specificity of the South.

Decentering the Site of Power

This chapter has called into question the conventional wisdom about union strategy, including the strategic planning tool known as power mapping. While undeniably beneficial in the case of subcontractors, simplistic power maps can, at times, obscure the many prongs of power. A power map seeks to locate power or identify its holder but rarely aims to interrogate the dynamic workings of power-as-process. The place of power is also a dis(place)-ment, an unmooring that defies the logic of position. Put differently, this chapter suggests that researchers must not overstate the impact of policy creation at the headquarters level. As with many GFAs, VW's "Declaration on Social Rights and Industrial Relationships" bore the imprint of its German signatories. Transnational union strategy sometimes remains wedded to a hopelessly outdated and largely obsolete understanding of the firm as an entity with particular characteristics linked to its headquarters. When VW's American unit began to betray corporate directives, unionists were left with little recourse, given their singular focus on headquarters as the site of power.

Indeed, efforts to forge connections across vast geographic distances simply because those workers share the same top-level employer may be a poor use of resources. Without genuine global governance, any global policy will necessarily take place in smaller arenas. For example, by virtue of its internal composition, the GWC's headquarters in Wolfsburg is the key site where VW's charter "takes place," its claim to global reach notwithstanding. Yet paradoxically, VW left enforcement of the GFA to local actors in Chattanooga, as predicted by Nikolaus Hammer and Lene Riisgaard.[74]

Affiliatory Control

While the particularities of this case make generalization challenging, it is not uncommon for GWCs to receive preferential treatment over GUFs as they, like GFAs themselves, operate at the level of the firm (rather than on an industrial or craft basis, as in the labor movement).[75] Elsewhere, it has been proposed that works councils are complicit in achieving worker consent to interplant whipsawing,[76] but our argument goes a step further, suggesting that employee representative bodies (including GWCs) may themselves be subjected to whipsaw-style competitive pressures. The profusion of points of contact between firms and labor-side interlocutors creates perverse incentives for interorganizational whipsawing ("solidarity arbitrage"), in which management exploits existing divisions and rivalries between various employee representative bodies to drive down expectations and escape strict regulation. Crucially, this process is often less explicit than conventional whipsawing, as management has largely retained the prerogative to select its own de facto labor designee from among potential contenders. Moreover, the perceived necessity for labor organizations to present a united front may cause them to minimize internal differences in public-facing forums and conceal competitive pressures.

Scholars have long recognized that scale can be weaponized against labor. They see preventing labor's ascension to higher scales as a primary way capital manages and diminishes labor's power.[77] By shunting worker organizations away from certain strategic partners and directing them toward others, capital exerts affiliatory control. This is "power acting upon power" (pace Foucault) to modulate and mediate solidarity rather than simply block or destroy it.

The literature typically has presented transnationalism positively as an alternative to the isolation and blinkered perspective that often characterizes labor. But connectivity is not an unalloyed good: the distinct properties of strategic actors (e.g., the GWC's firm-centeredness versus the GUF's industry basis) create perverse incentives.[78] Thus, corporations may strategically foreground weaker actors while sidelining stronger ones.

5

Solidarity Deferred

Why the UAW Lost (Again) in 2019

The previous four chapters, taken together, offer a comprehensive explanation of the UAW's defeat in 2014. The UAW encountered a formidable array of challenges (Chapters 1 and 2) for which it was underprepared and maladjusted (Chapters 3 and 4). This combination of strong
barriers to unionization and weak strategic capacity could only spell defeat.
Counterfactual speculation is risky, but it stands to reason that, by learning
from its mistakes in 2014, the UAW might have greatly enhanced its chances five years later.

Or so it seems. A detailed analysis of the 2019 election reveals that despite
fewer obstacles and a more sophisticated organizational response, the UAW
could not move the needle. Its margin of defeat was nearly identical to that of
2014. This raises the question, What went wrong? The defeat in 2019 suggests
that knowledge, motivation, and learning alone do not generate strategic capacity. Without the final dimension (innovation), organizations risk slipping
into an iterative feedback loop in which incremental course corrections substitute for more serious change.

This is the first of two comparative chapters that position the 2014 vote
against subsequent elections. This chapter covers the period from immediately after the 2014 election through the 2019 election, extending the analytical framework developed in Chapters 2 and 3 to the events of 2019. The
first section ("Interregnum") covers the period from 2014 through 2018,
when the UAW worked to make sense of its 2014 loss, experimented with
unconventional organizational forms, and developed an interim strategy it

hoped would lay the groundwork for the next vote. Equally important, the company was rocked by revelations that it had cheated on emissions tests, which tarnished the brand's image and led to corporate restructuring with significant ramifications in Chattanooga.

I then turn to the 2019 vote, which serves as a convenient comparative case because barriers to unionization trended lower while strategic capacity trended higher, allowing for a quasi-experimental design and an evaluative test of the strategic capacity model. The conditions that led to the defeat in 2014 had generally either disappeared or diminished somewhat by 2019, while the union improved its strategic capacity across metrics of knowledge, motivation, and learning. Specifically, 2019 was a much more vigorous campaign, with more meaningful worker involvement, a deeper commitment of organizing staff and resources, and a greater willingness to antagonize the employer. The UAW could not afford another high-profile defeat.

Interregnum: 2014–2019

Self-Organization

After its ignominious 2014 defeat, the union had a unique opportunity to start anew. Instead, as discussed in Chapter 3, the UAW tightened its operations and purged enemies from its ranks. If the union's conduct earlier had raised doubts, watching its position harden after its stunning defeat was the final straw for many. Workers' call for transparency evolved into a full-throated demand for union democracy. Most workers did not have deeply held political convictions at the outset of the drive, nor were they affiliated with the sectarian Left. Their desire for union democracy stemmed not from abstract principles but from a growing sense that the union and the company were engaged in deliberations from which they had been excluded. Therefore, they demanded union democracy on a strictly pragmatic basis. By injecting the UAW with vibrant internal democracy, they hoped to revive and energize a union that had sunk into a pre-election stupor. One worker wrote at the time:

> I don't feel that the UAW is interested in any kind of democracy at this point, and they will continue to facilitate meetings in a top-down fashion. I feel we need to assert our own independence to win any union vote in the future.[1]

As they recovered from their 2014 defeat, workers debated several possibilities, including organizing an insurgent rank-and-file caucus, running an independent slate, or attempting to affiliate with a non-AFL-CIO union. Until this point, the organizing had occurred entirely under the purview of the UAW, and even the union's harshest critics accepted that the UAW had a legitimate claim to exclusivity. Only after the 2014 drive had failed did workers begin to organize autonomously.

After the 2014 loss, workers exhibited a surprisingly high level of self-organization, particularly since few had prior experience as union members and virtually none had experience as participants in a labor organizing drive. While the high degree of behind-the-scenes union-company coordination might seem to preclude employee self-organization, the opposite was true. Relying entirely on their own initiative, worker activists conceived of a workers-only Facebook group that they used extensively for sharing, evaluating, and sometimes critiquing union strategy. Without timely, accurate, and complete information sharing by UAW staffers, the closed Facebook group served as a virtual union hall through much of the drive. This had certain advantages, permitting organizers to express themselves freely without fear of retribution. In the context of a drive that was tightly controlled and stage-managed by staff, the messy cacophony of social media provided a necessary corrective.

This private Facebook group served as a raucous open forum, giving rise to a range of views and suggesting the presence of a latent energetic undercurrent within an otherwise sclerotic organizing effort. Other Facebook groups figured prominently in the campaign; these included an official forum overseen by the UAW and a union opposition group under the banner No2UAW, but both operated essentially as unidirectional broadcast media. In contrast, the Facebook group was uncensored, lightly moderated, and inclusive. In a campaign that otherwise failed to inspire worker-organizers, this Facebook group stands out as a site of significant activity. In its most cohesive moments, the Facebook group may have been the closest workers came to achieving a community of shared interests and taking ownership of the unionization campaign. At other points, it was a freewheeling sideshow, dealing in vulgar, bawdy humor and hyperbolic rants as befit the medium. But beyond the back chatter, the group reveals the potential for self-organization, providing a locus of dissent and serving as a crucible of working-class self-activity.[2]

Social movement scholars disagree about the merits of the internet as a tool for organizing. For techno-optimists, the internet is supposed to facilitate a radical decentering of leadership.[3] Through its inherently democratic internal logic, the internet lowers barriers to access and transcends distance, increasing communication speed exponentially.[4] As activists engage in cacophonous online debates over movement strategy, their process-centered discussions resemble the idealized Habermasian public sphere. Moreover, new social movements are said to mimic the network-based format of the internet, deriving their very structure from the modes of communication they rely on.[5]

The Facebook group in Chattanooga resembles this description as it expanded and amplified the organizing campaign's reach. The group was considerably larger than the official organizing committee and broadened the UAW's support base. Through the internet, the unionization campaign attracted supporters from beyond the core circle of designated leaders. The volume of posts was tremendous, totaling tens of thousands over the group's year of peak activity, reaching over two hundred per day during moments of maximal engagement. The internet allowed gripes and grievances that otherwise might have been reduced to a hushed comment in a hallway to reach a mass audience. Posters drew on a far-reaching vocabulary, trading in memes, GIFs, and the detritus of internet culture while also linking to globally sourced press accounts. While most posts to the Facebook group dealt with the minutiae of everyday organizing, a significant number consisted of statements of support and mutual aid, suggesting that the drive engendered a sense of camaraderie beyond its immediate purpose.

While organizing wildcat strikes in West Virginia and Oklahoma, technologically savvy teachers relied heavily on Facebook to build organizational breadth and depth beyond the capacity of their unions.[6] Moreover, the internet can counter undemocratic business unionism, "defying hierarchical strategic direction and facilitat[ing] direct contact between worker geographical locations."[7] Furthermore, privacy features available on social media offer the possibility of clandestine organizing while providing activists with the cover necessary to speak their minds.

Yet there are definite limits to internet-based organizing. Some scholars have argued that the internet sets an artificially low bar for movement participation.[8] Though the internet can aid the recruitment of supporters by decreasing perceived risks, this same dynamic can encourage a kind of slacktivism, in which people with little commitment to a cause nonetheless feign involvement. This claim, too, finds some support in Chattanooga. While the internet allowed for *quantitatively* more voices, discussion, and expression, it did not *qualitatively* shift the balance of power. The internet seemed capable of broadening the reach and appeal of the campaign, but it did not create the deep connections necessary for a "culture of solidarity" to emerge.[9] Through internet activism, workers could plausibly identify as unionists without the real risks or depth of commitment that union building entails. Still, the Facebook group reveals that workers were keenly aware of the importance of their actions on the national and international scene. They often shared and commented on news articles, especially those that depicted their struggle in an unfavorable or inaccurate manner. Ultimately, the internet permitted the campaign's *extensification*, not its *intensification*. In fact, at times, the internet seemed to operate at some disconnect from the situation on the ground, offering an outlet for scheming and dreaming without risk or consequence. In a more cynical reading, the internet may have served as a release valve, diverting energy while giving the impression of action. For these reasons, overly optimistic evaluations of the internet as a site for activism must be tempered.

While initially promising, these efforts to build an independent organization never gained much traction among workers and never progressed beyond idle internet-based discussions. Ultimately, they faltered because of infighting, political disagreements, and organizer fatigue. Social media can intensify emotions and lends itself to the factionalism and partisanship that eventually overcame the organizers' best intentions.

For those not yet inured to the long slog through institutional power, it may be tempting to start anew and reinvent a bottom-up labor movement analogous to a guerrilla strategy that gradually wears down the enemy. However, guerrillas famously endure heavy losses, and the top-down strategy may still enjoy advantages against a heavyweight employer like Volkswagen. At this stage, most workers did not yearn to depose their leaders and self-manage the organizing drive on their terms. Instead, workers were clamoring for leadership that met Ganz's three criteria. They wanted leaders with access to relevant and appropriate information and the ability to communicate it clearly and readily. They desired leaders with the strategic competence and organizational backing necessary for a large and complex campaign. Finally, they sought leaders with enough sophistication, foresight, and humility to make changes as their best-laid plans inevitably went awry. Much to their chagrin, they got none of these.

Emissions Scandal

Immediately after the 2014 election, VW again found itself in the headlines, but for reasons removed from the union contest in Tennessee. Along with its troubles in Chattanooga, VW suddenly had much larger problems. The VW emissions scandal first came to light in September 2015. This corporate crisis would damage the brand's reputation and dominate VW-related headlines for several years to come. VW had installed in its vehicles electronic defeat devices that allowed them to meet emissions standards when undergoing testing, while emitting nitrogen oxide at levels up to forty times the legally permitted limit under normal driving conditions.[10] Defeat devices were installed in roughly 482,000 vehicles in the United States and some 11 million worldwide.[11] In June 2016, VW agreed to a $15 billion settlement in the United States, along with separate penalties from the EU.[12]

A month after the scandal broke, global VW sales fell 5.3 percent.[13] Though no Chattanooga-based executives were directly implicated, the events nonetheless affected the union drive in several ways. The scandal left a permanent stain on VW's Wolfsburg-centric leadership and cemented the firm's long transition from a centralized hierarchical multinational to a distributed transnational.[14] In the wake of the scandal, VW underwent a complete restructuring, which aimed to further decentralize its operations. After 2014, there was a deliberate push to distance Wolfsburg from local plant operations, resulting in more autonomy for national- and plant-level executives.[15]

Regarding raw production figures, Chattanooga's operations emerged from the scandal relatively unscathed. While U.S. sales of the VW brand fell 6.9 percent in 2015, the most popular models among American consumers were still produced overseas. Though dealers lost money and the brand was forever tainted, the Passat and Atlas lines (the only models assembled in Tennessee) were unaffected.[16]

Another consequence of the scandal, which had ramifications for Chattanooga, involved a shake-up on VW's board—one that seemed to portend a rosy future for the UAW. VW's most labor-friendly CEO in years, Berthold Huber, formerly of IGM, briefly took command of the company, ousting Martin Winterkorn, who had butted heads with established company patriarch Ferdinand Piëch. Unlike GWC chair Osterloh, Huber was on record backing the UAW. Yet Huber would be forced to balance competing interests in his new role, and UAW regional director Casteel was realistic about Huber's dual commitments, pointing out that he "[was] a good friend of the UAW no doubt, but he also [had] a fiduciary responsibility to the VW board."[17]

Local 42

Even as the scandal broke, the UAW, still reeling from the public embarrassment of its February 2014 loss, moved deliberately, if not unflinchingly, toward a strategy with little recent precedent on American soil or in the UAW's previous repertoire: minority unionism. On July 10, 2014, it launched "an innovative new VW workers' organization" with great fanfare. UAW president Dennis Williams commanded the stage as fifteen employees signed a charter to christen a members-only local, providing management with a union-like entity with which it could interact. Without providing much detail, Williams pitched Local 42 as an unconventional organization that harked back to the early, experimental days of autoworker organizing.[18] The company reacted with a tersely worded statement: "There is no contract or formal agreement with the UAW on this matter."[19]

Indeed, the UAW's turn toward minority unionism reflects growing interest in extralegal organizing frameworks. Under minority unionism, workers eschew the fraught NLRB election process in favor of unrecognized, informal unions that derive their power chiefly from aggressive and continuous shop-floor mobilization. Similar tactics have long been advocated by movement scholars like Staughton Lynd and fringe unions like the Industrial Workers of the World,[20] but have only recently begun to gain appeal among mainstream unions. It may have been desperation that drove the UAW to such measures, and their embrace of this tactic proved somewhat half-hearted. While the full potential of minority unionism was never fully realized, the fact that the UAW even considered this option is noteworthy.

Today, unions have become locked into the government-arbitrated labor relations system. Proposed remedies, such as the PRO Act (Protecting the Right to Organize) and EFCA (Employee Free Choice Act) before it, whatever their merits, only signal the limits of labor's imaginative horizon, as if labor's inability to build associational power and unwillingness to exercise structural power could be resolved through a legislative fix.[21] Given the growing ineffectiveness of the NLRB (a trend that slowed but did not reverse under President Biden), the idea that unions ought to seek representation outside the established parameters of labor law has gained traction in some circles.[22] The current model proved most effective during the era of mass-production industry, where employers defined the bargaining unit—an arrangement that has largely disappeared since the 1970s. But the present-day collective bargaining regime's calcified state obscures a deep tradition of extralegal unionism in the first half of the twentieth century. U.S. labor law extends many basic rights to workers in nonmajority unions, including the freedom to engage in concerted activity, the right to strike, and the ability to bargain collectively. Harking back to the bargaining style of the early AFL, some argue labor unions should again engage employers outside the stilted collective bargaining regime, relying instead on pressure politics and extracontractual campaigns.[23]

The UAW pursued both minority unionism and other forms of extracontractual negotiation in the aftermath of its 2014 defeat. But these efforts amounted to little more than timid experiments with minimal results. To be sure, Local 42 went through the motions of organization building, writing bylaws, designating leaders, and holding annual elections, but it ran no campaigns and mobilized around few issues. But the UAW never committed fully to this minority union experiment; instead, Local 42 was little more than a stopgap measure that allowed the UAW to buy time and maintain a presence in the plant as it awaited another chance at winning a standard (majority) union. As Chris Brooks wrote, "The group [was not] 'acting like a union' in any of the most basic, recognizable forms."[24]

Expectations for Local 42's innovative members-only model had been high,[25] but without a clear strategic mandate or a collective bargaining agreement, it was essentially dormant, excepting the sporadic Sportsmanship Night or community service event. Although Local 42 presented itself as a minority union, it was never more than an interim placeholder pending recognition and collective bargaining. The UAW was less interested in exploring the possibilities inherent in minority unionism than in laying the groundwork for the long-term goal of majority unionism.[26] Labor scholars have long been interested in mobilization strategies that circumvent the NLRB. Solidarity unions can serve as union training camps, preparing members to flex the solidaristic muscles necessary to build a recognized bargaining unit. They can also secure incremental wins during the interim by leveraging the power of collective action. But Local 42 did not serve either purpose because it had no organized presence on the shop floor. When members had grievances, they were more likely to call the company's HR hotline, despite its well-known ineffectiveness, than contact the local that supposedly represented them. It did little apart from social events, and international staff shot down efforts to mobilize around issues that might prove confrontational.

Microunit

If the UAW could not organize the most union-friendly automaker on the planet, its fate was sealed. Needing to rebound quickly from its ignominious defeat, the UAW turned to a strategy known as microunionism that exploits a loophole in the NLRA by allowing unions to organize small bargaining units when a wall-to-wall CIO-style organization is impossible. Local 42 president Mike Cantrell allowed that the UAW's eventual goal was to organize the entire shop but framed the machinists-only unit as a "step in the right direction."[27] Far from a revote, this new election involved a much smaller unit, including only the 164 skilled trades employees at the plant.[28] But by organizing a subunit consisting only of machinists, the union exacerbated a tendency that had created problems in the past: the domination and overrepresentation of skilled workers at the expense of production employees.

After initially suggesting it would bargain in good faith with this microunit, VW again reversed itself. The betrayal this time came fast and furious. As soon as UAW organizers were deployed, VW went on the offensive, claiming that the UAW's proposed bargaining unit fell short of requirements.[29] So-called microunits are complex from a legal perspective, especially when a larger group has already been defined as a shared community of interests. VW's legal filings undermined the claim that maintenance workers constituted a separate bargaining unit by playing down their degree of specialization. A VW spokesperson took to the airwaves to point out similarities between production and maintenance employees, noting that the two groups worked side by side on the floor, shared a bonus plan, and often spent time together in common areas like the café and gym.[30] However, the NLRB initially sided with the UAW and allowed the vote to proceed.[31]

Compared to the international showdown that had marked the 2014 election, there was little drama this time. It was well known that machinists overwhelmingly supported the UAW, so the UAW essentially opted not to campaign at all.[32] Though the union was understandably reluctant to deploy its full arsenal of tactics when a victory was all but preordained, mobilizing around public issues or grievances may have been effective as an organizing tactic. Meeting expectations, the election resulted in a decisive victory for the UAW with a 71 percent margin: 108–44.[33]

The nontraditional bargaining unit gave VW a legitimate basis to challenge the election and disregard its obligation to bargain collectively with the workers during the interim. A Donald Trump–appointed NLRB that barely concealed its hostility to microunits would dim the prospects for these workers. As the NLRB appeal wound its way through the courts, the UAW attempted to put a positive spin on a situation that bore all the hallmarks of an impasse. Even though the UAW's relationship with VW had, by this point, devolved into outright hostility, Casteel maintained his characteristically cautious approach when describing VW's labor policies. Desperately trying to spin an increasingly dire situation, he said, "They're not opposed to unionization per se. We're still talking to them."[34]

From the standpoint of workers in Chattanooga, their incipient microunit may have been the scandal's first real casualty. Evoking the scandal, Haslam questioned the timing of the microunit election, which he feared might amount to kicking Chattanooga's biggest corporate sponsor while it was already down: "Obviously, VW is struggling with many issues right now. It feels like to me the right time is for everybody to focus on addressing issues that VW has had and then turning around to producing great cars."[35] Sebastian Patta, the head of VW's Chattanooga plant, followed Haslam's lead, saying in a letter to employees that the UAW's timing for a new election at the factory was "unfortunate."[36]

VW almost immediately appealed the election results, presaging the 2017 *Specialty Healthcare* decision, which significantly expanded the criteria for determining the appropriateness of a bargaining unit.[37] The UAW, in turn, filed a counterclaim against VW, alleging refusal to bargain.[38] In an initial ruling, the NLRB sided with the UAW,[39] but VW refused to bargain, and the NLRB filed an unfair labor practice complaint against VW at the UAW's request.[40] Two years later, as the case lingered in NLRB purgatory, Trump-appointed NLRB partially rescinded the broader guidelines governing microunits, casting doubt on the UAW's ability to prevail.[41] From an organizational perspective, 2016–2018 were slow years at the plant. The UAW maintained that it was pursuing its organizing drive behind the scenes, but there were few signs of overt activity.

Having examined the UAW's brief and tentative experiments with minority unionism and microunits, we now turn to 2019, when the union mounted its second attempt at a plant-wide election.

Obstacles

Delay

In the spring of 2019, the UAW asked the labor board to call another plant-wide vote. Having suffered through an eighteen-month election cycle in 2014, the union aimed for a snap election, hoping to deny the opposition time to mobilize. However, for reasons beyond its control, this strategy backfired.

Although VW had long held the position that it would only permit a full-plant election, it proceeded to maneuver to block the revote, claiming that the still-pending issue of the machinists' microunit rendered the overlapping wall-to-wall unit moot.[42] The irony of VW using the machinists' unit—which it had long sought to discredit—to block a plant-wide unit it had consistently favored was not lost on UAW's lawyers. One dissenting board member noted that the ruling gave unions few options: "'Heads, the employer wins; tails, the union loses' cannot be the Board's new motto."[43] As a union press release stated:

> VW's current maneuvering to avoid an election in a wall-to-wall unit that it has for years argued is the only appropriate unit at its . . . assembly plant is disgraceful.[44]

After some hesitation, the UAW responded by disclaiming its previous win, seemingly realizing it had little to lose by forfeiting a unit that had yet to gain formal recognition.[45] However, the damage had already been done: multiple NLRB hearings and challenges resulted in the UAW's election being pushed from early May to mid-June 2019, leaving enough time for opponents to mobilize a formidable countercampaign.[46]

On June 28, 2019, the NLRB formally certified the vote, with 833 voting against representation and 776 in favor. Suffering another defeat by nearly the same margins seemed the ultimate humiliation. However, the UAW's spokespersons attempted a positive spin:

> Clearly VW was able to delay bargaining with maintenance [workers] and ultimately this vote among all production and maintenance workers through legal games until they could undermine the vote.[47]

Compared to 2014, shifting the blame onto the company made more sense. The plodding pace of the buildup to 2014 resulted from the UAW's protracted dalliance with card check, and once the union finally called for a conventional vote, the opposition was already fully mobilized. In 2019, the UAW had mounted a nearly clandestine campaign, then called for a snap election that caught many supporters—never mind opponents—by surprise. But six weeks of foot-dragging by the company granted anti-union forces sufficient time to mobilize and mount an oppositional campaign that rivaled that of 2014. Were it not for the ensuing legal battle—intended purely to delay the inevitable—the vote might have occurred without any organized opposition.

Honeymoon Effect

Some of the most serious obstacles from 2014 had become nonfactors by 2019, and others had diminished in importance. Specifically, by 2019, VW's early sheen had worn off. Although it had made significant investments in the community, it had failed to win over the city. No longer seduced by the company's boosterism, workers saw VW as just another car company, drawing direct comparisons to Japanese transplants. If the extended honeymoon lasted through 2014, by 2019, any lingering affection for their employer had dissipated. True to form, VW had accommodated itself to the American production system, shedding its German corporate culture in the process. In 2019, the UAW encountered a plant run not by mealymouthed social democrats but by stolidly anti-union executives, many culled from Japanese transplants. As before, these managers maintained a paper commitment to neutrality, but their actions and other public statements were so far removed from this promise that nobody seemed to take the pretense of neutrality seriously. By 2019, the company's American subsidiary had revealed itself as the union-busting firm many had suspected it was. Workers quickly overcame their initial infatuation with the company as the promise of a gracious new corporate benefactor wore thin.

Longtime workers consistently pointed to a drop in quality and morale as German supervisors cycled out, replaced by American managers conditioned by Japanese transplants. The honeymoon period had given the company a free pass for its first year of production, and this sentiment carried over through the 2014 election. But by 2019, any lingering goodwill had faded, as American midlevel managers took over the plant's day-to-day operations, and the company's relative largesse dried up.[48]

Worker attitudes also shifted alongside the economic cycle. The plant had opened in 2011, at the tail end of the economic crisis, which disproportionately affected automakers. Deflated wages and two rounds of concessionary bargaining drove down pay industry-wide. Politicians steered workers toward accepting the meager pay rate, not wanting a single company to sabotage Chattanooga's competitive business environment by driving up wages. By 2019, the industry had recovered, raising expectations for what workers might win.

Political Interference

As in 2014, political interference was a significant obstacle. The coup de grâce came in the form of a pep talk by a special guest of honor. Governor Lee was determined to fill the vacuum left by Corker's surprise exit from politics several years earlier. To reprise Corker's last-minute appearance in 2014, Lee's office coordinated with plant management to arrange a visit during the peak of the morning shift. Goaded with complimentary Chik-fil-A sandwiches, workers were herded into a cavernous auditorium and subjected to a captive audience session delivered by the highest-ranking elected official in their state. Though a company spokesperson said the visit was not union related, a clandestine audio recording of his speech contradicted that claim. Lee declared:

> It is more difficult to recruit companies to states with higher levels of organized [labor] activity. That is why, I think, it is in the best interest of the workers at VW—and really for the economics of our state— that VW stays a merit [nonunion] shop.[49]

Not long ago, VW had preferred to shield its dealings with the State of Tennessee from public scrutiny. By inviting the governor to address employees on company time, VW implicitly endorsed his message. Even by the standards of other transplants, this was a brazen moment of public-private confluence. A surreptitious record of the governor's speech includes cheers and boos, reflecting the divisions within the workforce.

Additionally, much to the chagrin of the UAW, Governor Lee (playing a role like that of Corker in 2014) was given full access to the plant and allowed to conduct a captive audience meeting in which he insinuated that the plant might close if it were unionized. Lee might have been more persuasive than Corker, whose threats in 2014 were delivered to the public via the media, not to a group of workers confined to a conference room at the employer's behest. Moreover, the exclusive access provision—a major source of conflict in 2014—no longer held sway. Instead, opposition forces were allowed to agitate inside the plant, while UAW staff were excluded.

As legal wrangling drew out the process, several familiar characters reemerged. Southern Momentum made a second appearance with the same lawyer/spokesperson but a new roster of rank and filers.[50] Southern Momentum had maintained a presence at the plant since 2014, but its activity was mainly limited to legal filings. Immediately after the 2014 election, Southern Momentum raised an alarm that VW might disregard the wishes of its employees and somehow install a union over their objections. While these concerns were entirely without merit, they reveal the depths of Southern Momentum's paranoia about VW and the UAW's supposed "collusion."[51] Anti-union advocates also rushed to file a lawsuit accusing VW of illegally conspiring with the union, arguing that VW had provided the names of workers and exclusive access to the plant in a quid pro quo arrangement that would require the UAW to maintain its regional pay differential in the event of a victory.[52] The Center for Worker Freedom also warned that the cooperation between the UAW and IGM, if financially motivated, may have triggered disclosure requirements.[53]

Union Substitution

The works council's role was diminished in 2019, but a different employee representation scheme now took center stage. First announced in November 2014, the Community Organization Engagement (COE) policy's appeal lay in its superficial resemblance to a form of union recognition. It provided workers with a scaled-back version of codetermination through high-level meetings with management and a taste of unionization, offering the right to discuss problems of mutual concern through progressively increasing levels of company access. (Notably, tiered access was capped at 45 percent, relieving the company of the charge that it had denied recognition to a majority unit.) However, it did not include provisions for grievance handling or collective bargaining. Finally, it was carefully crafted to steer clear of clauses in U.S. labor law that prevent companies from establishing their own labor organizations. As the COE document clarifies:

> This policy may not be used by any group or organization to claim or request recognition as the exclusive representative of any group of employees for the purposes of collective bargaining.[54]

Both the UAW and Southern Momentum attempted to work within the COE framework. Southern Momentum spawned a new group called the American Council of Employees (ACE). By its lawyer's own admission, ACE was an "anti-union union" established to compete with the UAW and erode its support; it did not aspire to become a bona fide collective bargaining agent. Instead, ACE hoped to bypass the messy business of unionization and move directly toward establishing a works council in Chattanooga. This strategy did not garner much support. The UAW rose to the top rung of the participation initiative almost immediately, quickly winning the support of 45 percent of the plant's rank-and-file workforce.[55] In contrast, ACE failed to rise above the bottom-tier threshold, with only 381 members at its height.[56]

The UAW's willingness to operate within the COE framework might be understood as a carefully calibrated strategy to preserve whatever traces of goodwill remained. On its face, the COE was a bold experiment. It is highly unusual for a U.S. company to engage in any way with minority employee organizations, particularly in the absence of militant action. Yet, in practice, the COE did not deliver meaningful gains to VW workers; instead, it offered the appearance of real negotiation without meaningful substance and may have even undermined later organizing efforts. To its critics, the COE represented VW's effort to save face and preserve its reputation despite management's unwillingness to deal with the union in good faith.

Frank Patta and UAW staffers spoke overwhelmingly positively about the COE while carefully acknowledging that it could never substitute for true collective bargaining. The COE was an exceedingly poor substitute for genuine representation. When pressed, unionists struggled to describe specific gains that had resulted from the policy. For many critics, the COE was little more than a consolation prize bestowed on a defeated union. Several went a step further, claiming that participation in the COE was not only useless but actually detrimental to the UAW in the sense that it diverted attention from VW's refusal to bargain with the machinists and indirectly served to legitimize the anti-union ACE, which was granted a seat at the table under COE, albeit in a diminished capacity.

Improvements

As in 2014, to stave off the union threat, the company announced a series of improvements timed to coincide with the election. These included reductions in overtime, a fifty-cent-per-hour raise, a boost in starting pay, and an increase in the maximum hourly pay-for-performance bonus.[57] In another classic move, management sought to alleviate some of the most galling concerns just before the vote, hoping to buy some goodwill until after the election. For example, they turned on chillers to moderate the temperature, modified the wardrobe policy to allow shorts, adjusted the weekly shift schedule, and kicked out an unpopular plant director. Just before the vote, VW had granted an across-the-board fifty-cent raise, bringing pay from $15.50 to $16—a paltry amount, just enough to allow the company to say it had increased compensation without "outside pressure." (Management was gambling on the theory that workers would not bother to make the comparison to far more generous Big Three contracts.)

At this point, VW employed one of the oldest tricks in the book with great success. Within weeks of the vote, highly popular former plant CEO Fischer (who had served from the plant's inception through the summer of 2014) replaced the reviled Antonio Pinto as the new face of VW's operations.[58] In the plant's early days, Fischer had commanded VW with the air of a benevolent landlord. Just as Lee and Corker, both developers with vast landholding interests, saw Chattanooga through the patronizing beneficence of the landowning class, Fischer seemed to embody all that was good about VW in his genuine concern for employees. Those with a sharp memory might recall that things had been far from perfect under Fischer, but his legend grew in his absence. By 2019, after half a decade of mismanagement by supervisors on loan from Japanese transplants, Fischer's triumphant return was treated as a second coming, as if he was on hand to mete out justice and rid the temple of thieves and grifters. In truth, Fischer was in Chattanooga to break the union—a point he conceded. The words that came out of his mouth to a captive audience in 2019 were nothing workers had not already heard, but intoned in his thick, German accent, they registered differently. If anyone could put a friendly face on the company's union busting, it was Fischer.

Disinvestment

Though the tone and tenor of the 2019 election were markedly different from those of 2014, the structural situation on the ground bore an uncanny resemblance. In another repeat of 2014, layoffs of temps preceded the vote. Even if direct-hire jobs were never threatened, the dismissal of temps created a sense of insecurity that wafted through the factory, and the election occurred beneath a cloud of unease.[59] Moreover, just as the 2014 election occurred against the backdrop of a long-awaited plant expansion, by 2019, VW was again on the verge of a new production line—in this case, an $800 million expansion (buttressed by $50 million in new state incentives) to produce an all-electric SUV by 2022, with one thousand new hires expected.[60] While this proposed expansion was less explicitly tied to the vote than in 2014, the subtext was difficult to ignore.

Just months after the 2014 vote, VW had unveiled a separate, $900 million expansion that would create two thousand more jobs to make a new midsize sport utility vehicle,[61] deepening the company's commitment to Chattanooga. The expansion announcement seemed to vindicate Haslam and Corker's argument that the company's success hinged on its remaining union-free. Predictably, they ran a victory lap, claiming that the no vote had delivered the new line. Haslam doubled down, continuing to harp on the possibility of job loss, complaining loudly in the press that other companies were beginning to shun Tennessee because of its problems with unions, and even conjuring a story—possibly apocryphal—of an employer who had rejected Tennessee because of the uneasy labor climate.[62] Whatever its merit, fear of job loss continued to supply the state's political class with a steady stream of invective, which lasted through the next two election cycles.

But the UAW was also quick to take credit for VW's continuing investment in the region. Union staff claimed they had cleared the way for a new product line by voluntarily withdrawing NLRB complaints after the 2014 loss. This statement was out of character for a union that had long sought to protect workers against the threat of plant closure and had been a leading voice in favor of the Worker Adjustment and Retraining Notification (WARN) Act, which mandates a sixty-day notice and payments in the event of an unanticipated shutdown.

The specter of disinvestment was again invoked in the week leading up to this election with the news that the UAW had successfully organized a small independent supplier in North Carolina that served VW among other plants in the Southeast. Lee seized on the opportunity, again stoking fear that a spate of unionization might deter Tennessee's burgeoning auto industry and discourage future investment:

> The more suppliers who come, the more it helps OEMs (original equipment manufacturers). That helps suppliers come. It's a virtuous cycle. I'd hate to see the cycle interrupted in some way.[63]

Union Suppression

In 2014, floor-level managers spearheaded the union busting, allowing higher-ups to dismiss their activities as unsanctioned overreach by rogue employees. In 2019, the company's anti-union effort was more coordinated and deliberate. The plant's president directly stated his preference for remaining union-free, and the company produced a website stating the same.[64] Official literature bearing the company's stamp of approval listed four recently closed plants organized by the UAW and warned employees that they could lose access to a program that allowed them to lease VW cars at below market rates.[65] Moreover, whereas the union had been granted special access to workers in 2014, company representatives bestowed similar privileges on themselves in 2019 as VW lawyers were permitted to walk the floor and lobby workers.[66] With its union-tolerant image now thoroughly discredited and its professed "neutrality" revealed as the disingenuous posturing it always had been, the company had little to lose if it kept its thumb on the scale.

VW's about-face may have represented not a genuine reversal of company policy but rather a logical response to intense financial pressure. As noted earlier, the city's tax breaks and other subsidies for additional investment in the plant constituted a mix of positive and negative incentives: funds could be withheld if the company became too union friendly, while supplemental funds might be awarded if the company resisted unionization. When asked directly why VW continued to deny the UAW recognition of its microunit, a staffer blamed VW's bottom line: "$900 million. That's exactly why. They want more money out of the state."[67]

Knowledge

The UAW, having watched in dismay as VW's social democratic precepts washed up on the shoals of capitalism, was taken aback by the company's "betrayal" in 2014. In 2019, having come to grips with the true nature of its adversary, it would not be fooled again. But the record suggests the UAW ought to have been suspicious of the company's "union-friendly" posturing from the start, as VW was not the first European firm to adapt its business practices to match American norms. As one worker said of 2014, "I paint the UAW as the person in an abusive relationship. . . . Even though there's a clear pattern, they just kept hoping that this time will be different."[68]

By 2019, the UAW had accumulated relevant knowledge to develop a more sophisticated understanding of the company. It came to realize that the VW struggle, for all its purported novelty, actually shared key elements with previous UAW campaigns. Indeed, most of the defining characteristics of the VW fight in Chattanooga—except management's early support for the works council—echo previous campaigns. While management's position on the works council was indeed unusual, it ultimately amounted to a series of unproductive conversations and the circulation of speculative documents, some of which VW later renounced. Therefore, it is useful to deemphasize the specificity of the Chattanooga case and reflect instead on the many parallels to previous UAW organizing efforts.

In sum, as time passed, both the company and union, with varying degrees of success, tried to triangulate the other's behaviors and motives, even as they shifted. In its early dealings with management, the UAW had a limited and impressionistic understanding of VW's labor relations model. The 2014 election and intervening years had exposed the company's true nature, and by 2019, the UAW had a much better understanding of its adversary. Over the course of five years, the company transformed from ally to antagonist. If VW semicredibly presented itself as a socially responsible firm in 2014, its stance toward the union by 2019 was practically indistinguishable from that of conventional American corporations. In response, the UAW shifted its approach from partnership to confrontation.

Motivation

Many of the UAW's difficulties in 2014 can be traced to motivational deficiencies (as defined by Ganz). In 2019, several factors led to an organizing drive that showed more motivational promise. First and foremost, during the intervening years, workers had begun to self-organize, breathing life into a union movement that had until then assiduously avoided its rank and file. But these improvements were encumbered by broader problems, including a corruption scandal that broke at an inopportune moment and a disastrous strike that only underscored the UAW's organizational problems and structural weakness. Together, these factors contributed to a mixed scorecard on motivation.

Rank-and-File Power

As its microunit, minority union, and COE plans fizzled, the UAW struggled to recruit workers to its cause. But the early stirrings of a self-activated union movement were already right under its nose. The 2014 vote had revealed the impossibility of imposing solidarity from above through top-down social engineering. It was fitting, then, that the impetus for a renewed organizing push in 2019 came directly from the shop floor. Recognizing Local 42 as an empty placeholder and the COE as simulated "unionism-lite," workers took matters into their own hands.

The year 2019 opened with a rare shop-floor action on the part of aggrieved workers. At this moment—fleeting though it was—the organizing drive took on conflictual characteristics against a backdrop of insurgency and worker-initiated direct action. When the line speed was unexpectedly increased, workers brought their concern directly to HR, bypassing the supervisors and managers who would normally be the first line of defense: "We walked into their office about twelve deep to let them know what was going on," said one worker.[69] The line speed was immediately corrected, and the responsible supervisor apologized personally to all those affected. One worker described the situation:

> We decided that's our route from now on—if we need something, it's directly to [management]. About 50 of us marched upstairs and said, "No, we're not going to do this." They couldn't start the line back up 'cause there were so many people missing from the line.[70]

The upshot of the march on the boss was to displace the machinists-only Local 42 with unsanctioned, informal worker-to-worker networks. Local 42 maintained its website between 2015 and 2019, but in practical terms, it was no longer a going concern. The UAW had disavowed its NLRB complaints over Local 42 when it filed for the 2019 election. Since the issue had never been resolved, the court's invalidation of the microunit stood, and it remained unrecognized in the eyes of the law. But in some ways, Local 42 had already been supplanted. Production workers were no longer willing to allow machinists to lead the way, preferring to take matters into their own hands. The independent grassroots organizing that had begun after 2014 and culminated with the march on the boss was driven and led primarily by production workers. Machinists remained involved, and "Local 42" provided a convenient masthead for formal documents, but the impetus behind the 2019 campaign came from workers who would have been ineligible to join the machinists-only Local 42 in the first place. In other words, organizing in the plant after 2019 crosscut skills and classifications while decentering machinists, creating an opening for truly representative shop-floor committees.

The UAW's success in the election would depend on its ability to channel the raw energy of the rank and file into institutional form. Workers were learning the power of direct action, but short-term wins could only be fleeting without the backing of an organized body to secure their gains.

Negative Momentum

Social movement scholars have long observed that participants derive inspiration and encouragement from one another's successes. As we see in Chapters 6 and 7, movements spread when mimetic imitation goes viral via social contagion. But the concept of demonstration effects has more often been applied in a negative sense, referring to the knock-on consequences of movements that are stopped in their tracks. These negative demonstration effects were in play in 2019. Anti-union forces seized on the worst parts of the UAW's post-2008 contracts, including tiered pay and the top-tier wage freeze. This line of reasoning was understandably effective in Chattanooga, which was already reeling from the effects of the de facto second tier composed of temps.[71] No matter how much the UAW spent on its media blitz, its decades-long history of concessionary Big Three contracts served as free advertising for the opposition. (Though the UAW made some slow and steady gains in the 2010s among smaller parts distributors, along with bigger, significant wins out of their core jurisdiction at universities, these victories lacked real demonstration power.)

By 2019, the UAW's sales pitch was not especially attractive. Following the 2008 bailouts, the UAW reached a historic low point. Post-recession contracts during the Obama era were stopgap measures to stem the tide of divestment and sources of considerable embarrassment for the union's promoters. According to an analysis by Bloomberg, in twenty-one of thirty-six years since 1987, UAW contracts included only lump-sum payments (instead of a compoundable wage hike).[72] Because these contracts with Big Three automakers were weak, they were not showcased as part of the UAW's organizing pitch to VW workers. Perhaps sensing that low morale and general indifference in its ranks, coupled with a ten-year pattern of concessionary contracts, would fail to entice new sign-ups, the UAW did its best to downplay its track record when organizing at VW in 2019. Even when the local paper contributed a cost-benefit analysis in an attempt to predict the net value VW workers might derive from enlisting with the UAW, noticeably missing from this cold calculation were the noncompensatory benefits of unionism, including safety improvements and a grievance procedure, not to mention an enhanced sense of camaraderie and solidarity.

The UAW had cut its members raw deals dating back to at least 2008. As part of the 2008 restructuring, workers relinquished their right to strike and acceded to a two-tier system that would keep post-2008 hires at a lower rate. Predictably, what was pitched as a temporary measure to save the auto companies from ruin quickly expanded until the second tier accounted for nearly half of all Big Three employees. The UAW's 2015 contract opened a path for moving newer, or "in-progression," workers up to "traditional" employee pay, but only through a complex process with numerous carve-outs that undercut its promise.[73] The divide over two-tier was a main factor contributing to the AC's unpopularity. Through the two-tier era, leaders prioritized their legacy first-tier members, even as they knew that second-tier growth was chipping away at the relative security pre-2008 hires enjoyed.

Armed with this information, anti-union forces could paint the UAW in broad brushstrokes, invoking weak contracts to discredit the union. If workers drew any lesson from the concessionary period of the 2000s, it was only discouragement. As the lead organizer from District 9A noted:

> Auto workers who are not in unions looked at the Big Three and said, "It's no better than what we have right now." Giving up things like COLA [cost of living adjustment], the tiers . . . the abuse of temp work. All of these things were just . . . very similar to what we experienced [in transplants].[74]

Indeed, a somewhat cynical reading of the UAW's southern strategy might reasonably conclude that organizing campaigns among foreign automakers were a defensive struggle to protect comparatively higher wages at the Big Three. Foreign companies set the floor to which Detroit was steadily descending. Hoping to stave off further decline, the UAW sought to establish a foothold in the South, believing it could preserve what remained of the social contract and defend its legacy base against the continued erosion of its living standards. A desperate strike against GM in 2019 may have hurt the UAW's cause more than it helped. Cynics speculated that the real goal was to allow workers to blow off steam and wear down members' fighting spirit until the subpoverty $250-per-week strike pay drove them back to the table.[75] Despite some favorable press coverage, the strike had a resigned, uninspired vibe and did not generate anywhere near the enthusiasm of 2023. The enfeebled GM strike and the subsequent contract, delivering only a 6 percent raise over four years, were reason enough to doubt the union's prowess.

Finally, by 2019, there were signs that the UAW's repeated defeats in Chattanooga were producing their own adverse demonstration effects. At least one worker thought losing in 2014 had set back the UAW's efforts elsewhere in the South, including at a Hyundai plant in Alabama, where a union campaign had struggled to get off the ground.[76]

Corruption

Further undermining its cause, the UAW made news for all the wrong reasons as the election date approached. Undercover investigations revealed that then-current UAW chief Gary Jones had used union funds to finance elaborate vacations at the luxury resort town of Palm Springs, California, running up tabs at lavish hotels, restaurants, and golf courses, all while instructing his bookkeepers to pass off his recreational pursuits as "union business." But if the inquiries have centered on Jones, it seemed clear that malfeasance pervaded the upper echelons of the UAW's officialdom. Jones had more than his fair share of enablers at Solidarity House (the UAW's headquarters), but many of these leaders were beneficiaries of shoddily concealed embezzlement schemes, dozens of whom pleaded guilty amid scores of indictments.[77]

At the time of the 2019 vote, the corruption scandal had not fully broken, and it was unclear just how deep the malfeasance ran. Still, the union's reputation was thoroughly tarnished nationally, and it would take years to undo the damage. But ironically, the multilayered bureaucracy that allowed Jones and others to abscond with funds protected its far-flung locals from knowledge of this misconduct. In the eyes of the public (and potential members), Local 42 was guilty by association, but in practical terms, Local 42 had no direct connection to the scandal. If the UAW's saturation ad buys before the election evinced an organization with deep pockets, the resort-hopping executives from Detroit still operated at a considerable distance from Local 42's makeshift office in a cinder block International Brotherhood of Electrical Workers hall. Yet, as the 2014 election had shown, Detroit's problems have a way of complicating the ground war in Chattanooga, and there was no sign that 2019 would be any different.

Under questioning, UAW staff dismissed the idea that the corruption scandal might have influenced the vote in any way. VW workers, they surmised, were indifferent to the machinations of stuffed shirts in Detroit. But the UAW's ulterior motives were apparent to even the most naive observer. The specter of a dying Detroit had been used effectively against the UAW in 2014, and in 2019, as rumors of embezzlement tarnished the union's long-suffering image, it had even more reason to keep the focus local. As a national organization, the UAW had been sullied, and the scandal in Detroit could never be fully separated from the ground operation in Chattanooga. The billboards that had portrayed the UAW as a harbinger of postindustrial ruin in 2014 were now emblazoned with indictment notices.[78]

Corruption undoubtedly tarnished the UAW's image, but the effects were even more profound. The corruption involved the misuse of members' dues money, but even more glaringly, it undercut the UAW's power by guaranteeing concessions in exchange for cash.[79] Not only had the union stolen dues, but it had directly undercut its ability to perform its basic statutory function. However, it is important not to overstate the impact of the criminal charges. The reputational damage inflicted on the UAW by the scandal was undoubtedly more serious than the monetary damage from stolen funds. For all its ripped-from-the-tabloids luster, the lavish vacations and spending sprees accounted for an insignificant portion of the UAW's overall budget. Even the headline-grabbing "racketeering" charge met the legal definition of the term on a technicality; there was never any sign of organized crime syndicates in the modern UAW. This was not a scheme on the scale of Enron or Bernie Madoff, where entire classes were defrauded of their life savings. Spread across the UAW, the total amount of misspent and improperly accounted funds was .01 percent of the UAW's annual budget, or roughly ten dollars per member. But more pointedly, corruption symbolized an antidemocratic culture that gave the upper strata free rein to conduct the UAW's affairs without the knowledge or consent of their charges. Even before the attorney general exposed the offending leaders, the graft was hiding in plain sight.[80]

Learning

The UAW earns a split decision on learning. On the one hand, it gleaned important lessons from the (negative) example of 2014, reviving best practices and returning to organizing fundamentals that had eluded it five years prior. In changing its approach, the UAW implicitly recognized its earlier shortcomings. At a minimum, the UAW seemed to have realized VW was not a reliable partner in employee relations and chose to deal with it accordingly. On the other hand, it was dumbstruck by superficial and ephemeral "improvements" that should have been met with a rapid response and counterattack.

The UAW's second plant-wide defeat was a demoralizing loss because, at least on paper, it was a vast improvement over 2014. Newly elected president Dennis Williams kicked off the drive with a promise to correct his predecessor's mistakes, insisting that he had learned from failure: "We learned a great deal of lessons about being more cautious about who is out there trying to undermine the process. We'll be more prepared."[81] As promised, the UAW did not commit the unforced errors that had contributed to its poor showing five years prior. Instead, it ran a campaign that generally met expectations, checked the boxes, and satisfied the minimal criteria for a viable organizing drive. There were meaningful improvements over 2014, especially in terms of staffing, messaging, and presumably, expenditures on items other than legal fees. However, in terms of member engagement, the union once again fell short of its mark.

Militancy

Recognizing that a soft-pedal approach had failed to deliver, the UAW emerged in 2019 with renewed ambition and tenacity. Most significantly, the union was under no illusions that VW would maintain its paper commitment to neutrality. Its campaign seemed ripped from the pages of the Organizing Institute training program, ultimately enlisting forty organizers to staff its makeshift campaign office (roughly one for every forty-five bargaining-unit-eligible workers, or ten times the number in the earlier drive). Since there was no neutrality agreement this time, the union was free to make house calls and took full advantage of this opportunity. One organizer claimed that every worker was contacted at least once, and some several times.[82]

Compared to 2014, when its campaign lacked venom, the UAW ran a more aggressive operation five years later. Perhaps trying to mollify critics, the UAW revived the controversial house visits in 2019—a standard feature of most organizing drives but one it had pointedly avoided in 2014. But these visits were conducted mainly by professional staffers from out of town. Just as in 2014, this was a staff-driven campaign, except now, the staffers were knocking on doors rather than coordinating transatlantic summits. This also seemed to backfire, providing the opposition with evidence that outside agitators were seeking to influence the vote by infiltrating the sacred space of domestic life. Plant CEO Antonio Pinto stated that "employees [had] been victims of extremely distasteful bullying" and "employees [were] being followed to their homes and confronted about their views."[83]

In short, compared to the anemic campaign in 2014, almost anything would have registered as an improvement, but this was not the all-out, no-holds-barred approach one might expect from an organization that had staked its future on defeating what was then the ninth-largest company in the world by annual revenue.[84] Unfortunately, such superficial fixes could not resolve the more fundamental problems, some of which bubbled to the surface in 2019 through a series of ill-timed coincidences. Though the UAW earned high marks for learning in 2019, these alone were insufficient to rescue its organizing drive. To escape its past, the UAW would need to exceed the short-termism of quick-fix solutions and address the deep-seated problems that cut to the heart of its organizational identity.

Communications

In another break with past practice, the UAW seemed determined to outmaneuver the opposition on the public relations front. It mounted a major media offensive, spanning print, billboards, TV, and the internet. Though Chattanooga is a small market, the advertising came at no small expense. The union's campaign literature and publicity highlighted paid time off (PTO) and VW's unwillingness to provide a defined benefit pension plan.[85] As before, wages did not take center stage; instead, they were supplanted by scheduling and benefit-related issues. In contrast to 2014, when its broadcast presence was minimal, in 2019, the UAW circulated a series of professionally produced videos depicting VW in a negative light. Notably, they prominently featured workers who had voted no in 2014 but then changed their minds. Much of the expenditures seemed aimed at combating the anti-union campaign one-to-one, with $50,000 allocated for radio and television ads to match the opposition's aggressive air war.

Democracy

Despite these improvements, there were limits to the UAW's learning behavior. Even in 2019, the UAW was reluctant to encourage real democracy. For example, when a group of workers sought to use the Local 42 infrastructure to circulate a petition demanding a more consistent work schedule, UAW staffers rejected them. Local 42 eventually assembled a survey on the issue, but the results were not publicly shared and were not used as a mobilization tool. Reportedly, staffers were worried that independent organizing might be too successful, suggesting to workers that they could win on their own, without the union's help.

The organizing committee was roughly double the size of 2014 (sixty workers versus thirty for a plant that had increased by 10 percent), but it was still smaller than best practices dictate. Many of the union's actions had a perfunctory character, involving the organizing committee but few others. Flyering initiatives and T-shirt days attracted scant interest; even organizing committee members were unreliable participants. Among its more serious mistakes, the UAW made no real effort to turn out workers for the only pro-union community rally organized in the weeks leading up to the vote (a few showed up, but strictly of their own volition). One worker said, "There was no organization geared to bringing people together"—a harsh indictment of a campaign dedicated to that very purpose.[86]

Innovation

This chapter has shown that knowledge, motivation, and learning practices are necessary but insufficient waypoints en route to solidarity. Learning may produce short-term, incremental gains, but sustained improvement may require deeper changes that can only be achieved through innovation. Readers will recall that by 2019, the UAW had held half a dozen elections at southern transplants, plus additional aborted attempts that never achieved enough support to reach a vote. For longtime observers, each defeat was barely distinguishable from those that preceded it. The cycle of failure was propelled by a string of promises to "keep trying," perhaps with slight variations based on plant location, company-specific factors, or other strategic considerations but otherwise hewing to a predictable pattern that remained remarkably consistent over time. As research predicts, failure rapidly became cumulative (self-defeating) since each subsequent effort bore the mark of past defeats and was narratively positioned against a backdrop of futility. In its attempts to extricate itself from this pattern, the UAW struggled to innovate while contending with the inertial force of history.

Its brief experiments with microunit and minority unionism notwithstanding, the UAW essentially treated 2019 as a do-over of 2014: a chance to run an unremarkable, standard-issue union campaign without unforced errors. This allowed it to make a respectable showing despite much more intense employer hostility, abetted by the notorious anti-union law firm Littler Mendelson and a personally delivered anti-union speech from the governor. With a more sophisticated campaign strategy and active learning from past defeats, the UAW had a fighting chance against much stiffer odds.

In short, by learning from past defeats, the UAW avoided a humiliating blowout and lost by essentially the same margin as five years prior. Regrettably, union elections do not come with consolation prizes for almost-victories. Though the UAW made real improvements compared to the low bar it had set for itself in 2014, without innovation, it could not address broader problems endemic to the organization itself. Innovation was all but absent in 2019. I defer a more extensive discussion of innovation until Chapter 6, where I develop the concept in greater detail.

Conclusion

Up to a point, a learning perspective helps explain the union's changes from 2014 to 2019. The UAW emerged from 2014 savvier, more experienced, and far more cynical about the company it had once held in high esteem. VW, it turned out, was a typecast stock character, barely different from the Japanese automakers it had battled for thirty years. Ultimately, the UAW's loss in 2019 more closely resembled the pattern of early twenty-first-century defeats by anti-union employers in which even the most aggressive organizing playbook was often no match for the emboldened corporate opposition and its political enablers. The union emerged from 2014 having learned to combat hostility from shop-floor managers, distrust hollow pronouncements from management, maintain some strategic distance in its dealings with overseas executives, and avoid the distraction of elaborate works council schemes en route to unionization. It developed a longer institutional memory, regained the trust of its members, and improved communication. But divestment remained a persistent concern, which the opposition again mobilized to considerable impact. Additionally, despite a valiant effort, the union was unable to stave off the delay tactics that had hurt its chances in 2014.

In the final analysis, although the UAW ran a more inclusive and member-driven campaign, this shift ultimately came up against the hard limits of the organization's undemocratic character. Still hampered by its bureaucratic structure, the restrictions of U.S. labor law, weak transnational institutional mechanisms, and force of habit, the UAW's learning behavior could only go so far.

6

SOLIDARITY IN CONFLICT

Insurgency, Contention, and Union "Democracy"

Joseph Schumpeter's famous depiction of innovation and obsolescence under capitalism took the form of a compound term: *creative destruction*.[1] In standard grammatical usage, the adjective modifies the noun, qualifying the entity it describes through its attributive function. But intriguingly, Schumpeter's construction is ambiguous in tone, conjoining an affirmative descriptor (*creative*) to the deprecated word (*destruction*) that follows it. Destruction is the condition of possibility for the creativity it enables.

Crucially, I follow Schumpeter in reading innovation as both a destructive and creative act that breaks with the past as it conjures new futures. To break free from the past, the UAW would have to dissolve old ties and slough off obsolete habits in an ecstatic bacchanalia of destruction. Yet it would also need to impose a new order to carry it forward, channeling the pure negativity of its organizational death drive into an affirmative program for the future.

As noted in Chapter 5, improved strategic capacity was no match for the deep-set structures limiting the union's innovation ability. This chapter picks up where Chapter 5 left off by expanding on the concept of innovation, highlighting its dual character as a creative *and* destructive act (as per Schumpeter).

The first sections focus on innovation's destructive pole, showing how the UAW liberated itself from the spatial, relational, and organizational constraints that had held it back in 2014, dissolving old ties and sloughing off obsolete habits. In the remainder of the chapter, we turn toward the creative dimension of innovation, showing how the UAW remade itself in 2023, channeling the pure negativity of destruction into an affirmative program.

Organizational Failure

The dominant paradigm in organization studies has often presented failure as a surface-level phenomenon to be averted through surgical interventions. Organizations are said to stray from their objectives because of strategic miscalculations or tactical errors that can be identified, isolated, and fixed—as long as the problem is accurately understood and correctly diagnosed. In this view, failure is a surface-level blemish on the otherwise healthy organizational body, while recovery is a quick-fix band-aid. With the help of change agents, organizations traverse the gap between their original, flawed state and their final, remediated state by advancing linearly toward their goals. While this perspective may benefit organizations facing minor difficulties, it is poorly suited to addressing structural or endemic challenges. In these situations, organizations may be incapable of self-healing and risk devolving into cycles of repeated failure that become hardwired, leading to frustration or despair.

As noted in the Introduction, what Ganz calls "heuristic processes" constitute something of a catch-all master category that includes "understanding complexity, keeping response options open, suspending judgment, using wide categories, breaking out of 'scripts,' brainstorming, and playfulness with ideas."[2] Because this index is multidimensional, it may be difficult to operationalize. But learning practices should not be conflated with innovation. My strategic capacity model departs from Ganz in that it treats innovation as a semiautonomous factor, distinct from and irreducible to learning. Learning is limited in its analytical value because it draws on an incremental model of change in which organizations make minor iterative course corrections through a dynamic feedback loop.[3] By focusing on learning alone, organizations risk slipping into a cycle of infinite regress in which all future behavior is conditioned by the past and dismissing potentially innovative solutions simply because they do not correspond to previous experience. In contrast, innovative thought involves experimentation with new and untested strategies and a willingness to defy doxa and abandon habitual strategies that no longer produce returns.[4] Innovation must be treated as a separate analytical category, no longer tethered to learning, which reaches its limits when problems are deep seated, structural, or endemic, making self-correction through iterative change all but impossible.

Learning is also unlikely to address deep structures. If surface-level change is continuous and frequent, more fundamental transformations are rare because the organization is guided by the gravitation pull of a "highly durable underlying order" that limits the scope and magnitude of change and preserves institutional order.[5] These deep structures include tacit knowledge and taken-for-granted conventions that are deeply ingrained in the organizational fabric and may be expressed only through unconscious shared values. Nonetheless, they endure at the subterranean level to shape events on the surface level. Deep structures play an important role in shaping organizational change. Deep structures come to light particularly during Ganz's "focal moments," during which changes in strategic capacity open up to changes in leadership and organization. If this surface level is only a staging ground for changes influenced and constrained by deep structures, any programmatic change agenda must probe deeper and question the fundamental assumptions that maintain the system's stability. Some structures may seem so fundamental as to be beyond reproach; these, too, must be reevaluated.

Deep structures persist sight unseen but redound to the shop floor. To rid itself of the shadow of the past, the UAW would have to clean house, casting off deep structures and throwing open the door to an unwritten future. Here, I revisit those deep structures to examine the UAW's organizational unconscious circa 2014—the subterranean logics and unspoken assumptions that underpin organizational life and shape day-to-day affairs from behind the scenes.

Insurgency

UAWD

The UAW has a long and complex relationship with democracy. From its founding at the height of the Great Depression and through the war years, the UAW saw bitter infighting between two caucuses—one Communist, the other, which later became known as the AC, led by Walter Reuther and his disciples. These groups functioned as full-fledged political parties, each with its own organizational infrastructure and internal factions that further deepened the rich culture of democracy. Since internecine rivalry was such a defining feature of the UAW's early years, students of union democracy have often treated the union's first decade as an exemplary case study.

This changed in 1947 when Reuther finally purged the Communists and installed loyalists from his aptly named Administration Caucus (AC) in all top leadership positions. Reuther justified his autocratic turn by telling his convention that bargaining "requires central direction in terms of timing and strategy and tactics, and if we dilute this central direction . . . you dissipate the power of the union."[6] From 1947 through 2023, scholars of unions saw the UAW as the very model of entrenched union bureaucracy, even if its center-left political orientation sometimes gave it a free pass. Nonetheless, though the AC managed to preserve its grip on the UAW without any real opposition for seventy-six years, there have been sporadic efforts to topple the one-party state.

Throughout the seventy-six-year reign of the AC, rank-and-file movements challenged power from the outside, occasionally winning elections but more often running issue-based campaigns focused on achieving accountability. Though formal democracy was absent, the AC's one-party state harbored a strong undercurrent of dissent that periodically threatened to disrupt their control. For example, in the 1960s, Black militants organized the Revolutionary Union Movement at factories across southeastern Michigan, drawing inspiration from the Maoist "internal colonization" thesis to argue that Black people formed a hyperexploited subclass in both the workplace and the white-led labor movement. Labor leaders sold their labor power at a cut-rate discount, all while denying them proportional representation or restricting them to menial roles. Though Revolutionary Union Movement affiliates fielded candidates and occasionally won local elections, the movement's main focus was outside agitation. The Revolutionary Union Movement forced the UAW to confront its hypocrisy in styling itself as an ally of civil rights while tolerating de facto segregation on the factory floor and excluding Black workers from most influential staff positions.[7]

Another wave of resistance emerged in the 1980s; the New Directions caucus mounted a series of successful challenges to the AC across the Great Plains states, eventually electing one of its own, Jerry Tucker, to the UAW's executive board. New Directions had impeccable timing, emerging just as the companies were leaning on the union to accept lean production, cooperate with the outsourcing of parts plants, and permit stiffer penalties for absences, among other concessions.[8]

In the forty years since New Directions, opponents of the AC never gained much of a foothold. Forty years would be a remarkably long autoworker career, so all but a handful of New Directions activists had retired, but many offered informal advice or assisted in a consultatory capacity. But the UAW's rank-and-file rebellion had been brewing for some time before it coalesced into UAWD. Moreover, despite appearances, the AC was never a pure monolith; its officials presented a united front once in power, but the elections were often sites of acrimonious rivalry. Signs of discontent occasionally translated into outright rebellion, as in 2015, when Stellantis (then Chrysler) workers rejected their national contract, citing its two-tier provisions.[9] Yet while these contract fights signaled discontent, they expressed themselves through fleeting moments that never congealed into a sustained rebellion, much less a caucus with its own institutional power base. By the time the reawakening that would lead to UAWD had begun, organized opposition to the AC was all but moribund.

In its earliest attempts to grasp at power, UAWD was a tiny faction that enjoyed support from scattered locals but lacked the infrastructure or national reach to challenge the AC at the national convention, much less capture Solidarity House. But arrayed against an indifferent leadership and a membership inclined to cynicism, a militant minority can inject a much-needed sense of vibrancy into organizations that tend toward dejected diffidence.[10]

UAWD got an unexpected boost from the U.S. Attorney's Office, which ordered a vote on direct elections as part of a settlement package. Though the union's autocratic tendencies were not themselves criminal, the abuse of power by top officials was enabled by their unaccountability and shielded from scrutiny through the organization's general antidemocratic culture. In this sense, the high-level dealmaking that characterized the debacle at VW can be viewed as an outgrowth of the concentration of power in the hands of the officialdom, who saw it as their prerogative to impose a works council by fiat, whether or not such a strategy had any traction on the shop floor. Though the UAW's assigned federal monitor remains controversial, the crackdown on corruption ushered in a different culture that is more transparent, less vulnerable to abuse, and more easily overseen.

UAWD's first initiative was a vote on whether to allow a vote on the direct election. After fighting for a special convention to decide on direct democracy in 2020, an effort that fell short, UAWD saw its first win at the 2022 convention when the International Executive Board agreed to UAWD's proposal to raise strike pay from $275 to $400 per week.[11] After putting the question of direct democracy to the membership, direct elections won by 63 percent, despite low turnout (itself a consequence of disenfranchisement).

The AC held the incumbent's advantage and could argue that its experience lent it credibility, especially since the scandal was evidently limited to a select group of individuals at the top of the ballot. Moreover, the AC was deeply entrenched in the organization since even local staff were caucus appointees. A reformer had not won a seat on the UAW board since 1986, and the last serious challenge for the presidency had occurred in 1946.[12] The UAW had a rich tradition of appointing members to staff positions, so some staffers understandably felt a sense of loyalty to the leaders responsible for allowing them to ride out their days in the relative comfort of a climate-controlled office rather than grinding down their bodies on the factory floor. Under the AC, regional directors, chosen by convention delegates, are charged with keeping locals in line and keeping a lid on discontent. As one worker noted:

> The membership has been complacent because we didn't have anywhere to go. Every time somebody tried to do something, it just got shot down by the AC. They've just been beaten down into their place, and it kind of sucks.[13]

Importantly, though UAWD now controls the executive, its rise to power has been marked by contention, and its grip on power remains precarious. The UAWD was constantly guarding against counterreform from the AC. To sell its candidates, UAWD had to play to its base and differentiate itself from the old guard and other rivals. This competitive climate pushed UAWD toward an aggressive posture, as the organization adopted the uncompromising slogan "No Corruption, No Concessions, No Tiers." The UAWD could not afford to be complacent under these circumstances. Its push for reform was bitter and hard fought, with both caucus holdouts and other dissident factions resisting UAWD's initiatives and running alternative candidates. Fain was ultimately elected to office on a narrow margin after a contentious campaign whose outcome remained uncertain until the final votes were tallied.[14]

Organized resistance to Fain's leadership has never shied from seeking to disrupt his administration. At the 2022 convention, some UAWD-sponsored resolutions were booed, and Fain himself endured heckling from AC loyalists. In a show of protest, some old-guard regional directors initially refused to promote the contract campaign—behavior that would have been met with swift retribution from the AC. By the time of the strike, pressure from above and below had caused these regional directors to fall in line, but not before they had publicly registered their disagreement with the upstart caucus now in command of their union.[15]

Even once in power, UAWD took a remarkably different tack than its predecessor. Still haunted by the ghost of the Communist caucus, the AC had discouraged real debate, seeing the independent expression of viewpoints as an implicit threat to its party line. In contrast, UAWD encouraged and welcomed new ideas, recognizing that disagreements can only be sorted out through their free expression. Even as members debated the contract, UAWD went as far as to maintain an officially neutral stance on how members should vote (even though some of its highest-profile figures had sat on the bargaining committee and were now tasked with selling the deal to the rank and file). UAWD also organized speak-out meetings, creating a forum for debate outside the oversight of elected officials with split loyalties.[16]

Leadership

Accounts of the UAW's rank-and-file rebellion often focus on the singular figure of Fain, whose fiery oratory and polarizing rhetoric evoke the heroic sagas of labor's past glory. Profiles in popular media lean heavily on thoroughly discredited "great man" historical narratives, presenting Fain as commander, chief strategist, and lead propagandist, all packaged behind his disarming midwestern smile. Though theories of transformational leadership have long been out of favor in organization studies, they maintain their grip on the popular imagination. Under this model, an organization amid crisis is best served by a newcomer whose personal charisma and visionary fore-sight inspire a passionate following and mark a pathway toward collective salvation. The religious overtones are no accident here—as Fain seems to recognize with his frequent invocation of biblical references. A thorough evaluation of transformational leadership is beyond the scope of this book, but among the most salient criticisms is its inattention to contextual factors. The effectiveness of leadership can never be reduced to individual traits; it is the by-product of forces arrayed across the social field that shape both lead-ers and followers. In Fain's case, these contextual effects are particularly im-portant because his improbable ascent from midlevel functionary to presi-dent of the International was never assured.

If Fain's election is often seen as the triumphant culmination of UAWD's long struggle, his ascension to the presidency is only the most visible aspect of a prolonged and multipronged fight for democracy, as described in the previous section. The man who is today the public face of the UAW never had an easy path to victory. Fain may be the most prominent UAW leader. But his cabinet and top advisers are largely responsible for the pronouncements, speeches, and strategic innovations. In particular, Fain's inner circle includes three left critics of the UAW culled from the dissident publication *Labor Notes* and its associated activist milieu. Others followed a more traditional path, having risen through the ranks only to break with the AC and align themselves with Fain.

As Ganz writes, this heady combination of experienced perspectives and dissenting voices makes for good leadership.[17] Ganz insists that leadership benefits from vigorous internal debate and dissensus at the top organization-al levels (or deliberative processes). Within limits, the dialogical melding and sorting of ideas generates optimal solutions. Conversely, when leaders are surrounded by yes-men (and yes-women), party apparatchiks, or layers of ingrown officialdom, innovation and creativity are unlikely to emerge. Innova-tion necessarily involves risk-taking, especially in its early stages. Openness to innovation requires that organizations have some tolerance for failure, not-withstanding a certain number of inevitable disappointing outcomes. Think-ing outside the box through "deviant perspectives" facilitates better deci-sions and encourages innovation.[18]

For his part, Fain is now eager to flex his leftist bona fides, but his relation-ship with the rank-and-file caucus that fueled his ascent to power is not straightforward. Shawn Fain, more of a creature of habit than a maverick outsider, was deeply entwined in the UAW's bureaucracy by the time of his election. Fain was the handpicked candidate of UAWD, but he served under the AC until 2022 and was only peripherally involved with rank-and-file movements. The conviction of top UAW officials finally blew the door off the AC's tight ship, turning even party loyalists against the existing state of affairs.

Though the UAW's Detroit-based International Headquarters office had allowed Region 8 to handle the 2014 election at arm's length, top leaders were more deeply involved in 2024. Among Fain's top advisers was Chris Brooks, a Chattanooga native who first gained national prominence as a coordinator of community groups backing the 2014 campaign. His efforts were shunned by the UAW's leadership at the time. After spending years organizing teachers in the Midsouth and aligning himself with the labor-left publication *Labor Notes*, Brooks emerged as one of the more visible figures in UAWD and eventually won a seat on Fain's cabinet, serving as his chief of staff and right-hand man. Whatever Fain's personal investment in Chattanooga, Brooks coveted VW, and his southern twang and Christian principles lent him an air of legitimacy.

It was a happy accident that Fain had local roots: three of his four grandparents were from Tennessee.[19] The specter of Chattanooga hung heavy over Fain's candidacy. His opponents included Chuck Browning, a veteran of the 2014 organizing campaign and former director of Region 8, which includes Tennessee. Ganz notes that leaders whose personal histories connect them with multiple constituencies or transcend diverse populations can be valuable, as these people have divided loyalties that may translate into greater empathy. Fain was a former electrician who rose through the ranks to the top spot in his union. As such, he had experience as both a blue-collar and white-collar worker, holding both rank-and-file and staff positions.

Stand-Up Strike

The strike broke from established patterns in too many ways to describe. But this was symptomatic of an organization that had reinvented itself on the fly, moving from a one-party state to a vibrant organization with a contested internal life rife with dissensus.

Unpredictability

Though some criticized the union for failing to mount an all-out strike, Fain argued persuasively that a limited strike could be more effective. The stand-up strike had several key advantages: it hit the most valuable targets while rationing strike fund reserves, inflicting maximum damage at minimal cost. Fain indicated his strategic logic in an announcement to members:

> We know their pain points. We know their money makers. We know the plants they really don't want to see struck.[20]

As important, the strike provided incremental incentives, rewarding companies for progress and punishing intransigence. As negotiations unfolded, the companies often granted key concessions immediately before Fain's Friday deadline, suggesting that his strategy worked. Later, after initially escalating strikes only on Fridays, the UAW shifted to unannounced escalations on any day of the week.[21] Given the forty-year history of capital flight that had decimated the union's ranks, among the most significant wins was an agreement to allow the union to strike over plant closures.[22] The UAW called strikes without notice, alerting workers and management only minutes before a work stoppage began. (In practice, workers were sometimes tipped off slightly earlier.) Through this practice, the union kept management on its toes while keeping workers glued to weekly announcements, encouraging speculation on which plants might be next to strike.

Among the many limitations of American labor law is its strict regulation of permissible activities during job actions. Far beyond the time and place restrictions constraining free speech, the state restricts both the form and content of concerted activity by workers, prohibiting sympathy strikes, mass pickets, and hot cargo. Given the limited window of possibility, every job action bears a striking similarity to the next. However, the UAW broke from this pattern in two seemingly contradictory ways. On the one hand, its stand-up strike expanded the terrain of struggle, targeting three employers simultaneously rather than one at a time, as is customary.

On the other hand, the strike was focused and narrow in scope, selectively targeting individual job sites rather than all locations within the bargaining unit. In other words, the stand-up strike substituted leverage for density, making up in intensity for what it lacked in extensiveness. In doing so, it also reintroduced an element of surprise, which had been important during the sit-down strikes. Rather than signaling its intention to strike a plant in advance, the union withheld information until the last minute, preventing the company from stockpiling reserve parts.

Distribution Centers

As noted in Chapter 1, the devolution of parts suppliers has weakened autoworkers' structural power. As parts suppliers are spun off or outsourced, their relative power increases, creating codependencies throughout the production network. Capital worked to place suppliers beyond reach, separating their interests from those of mainline workers. Labor's struggles in the 1980s and 1990s centered on resisting outsourcing, playing into an us-versus-them logic that pits parts against assembly. Recently, the UAW had been quietly organizing parts suppliers with some success, but these victories were not strategically or rhetorically connected to the highly visible campaigns at major plants.

The struggle over parts distribution centers parallels that over parts suppliers. The Big Three operate about sixty after-sales parts distribution centers across twenty-five states, mostly close to major metropolitan regions since they serve dealerships and repair shops rather than manufacturing plants. OEMs have established new subsidiaries such as GM's Customer Care and Aftersales (CCA) unit and Stellantis's Mopar division. They are a "technological bottleneck" or strategic choke point on which the rest of the sector depends.[23] Lingering supply chain volatility since the pandemic has only boosted demand for aftermarket parts.[24]

Parts depots are historically beset by lower wages, worse working conditions, and more pervasive health and safety issues than mainline production. During the 2008 restructuring, even as UAW negotiators drew attention to incremental improvements, the rise of distribution centers and increasing reliance on temporary agencies were not simply imposed on unions from the outside but rather implemented with their tacit cooperation. The union contented itself with debating the terms of two-tier rather than contesting the model itself. Under the 2008 restructuring, company representatives presented spinning off parts depots as an inevitable conclusion on which the industry's ongoing viability depended. This framing foreclosed the possibility of a different future, narrowing the Overton window and compelling unions to content themselves with negotiating over distribution centers rather than questioning the model itself. Thus, by 2023, the UAW found itself in the unenviable position of striving to undo an arrangement in which it had been complicit. While the UAW has fought to reintegrate tiered, outsourced, subcontracted, and irregular workers into core contract provisions, the companies have sought to increase their reliance on nonstandard work. Heading into negotiations, Stellantis proposed closing ten Mopar distribution centers and consolidating them into one Amazon-style warehouse, with a predictable effect on employment numbers.[25]

Though parts *suppliers* could not themselves be struck under the terms of labor law, Fain pulled parts *distributors* even before all assembly plants had been shuttered. His actions testify to their outsize power. During the 2023 strike, even as most manufacturing plants continued production during the calibrated shutdown, the UAW stopped work at all GM and Stellantis parts distribution centers, totaling twenty-seven unique locations. So critical are these centers that Stellantis preemptively sought to staff up with nonunion temps in anticipation of a strike.[26] Pulling distribution workers allowed the strike to reach areas of the country beyond the UAW's base, including California, Texas, and the Southeast—all target regions for transplant and EV organizing. In this sense, distribution centers helped prime the pump for the next wave of organizing while also hitting consumers close to home. The strike was already dominating national headlines, but for news outlets in major markets outside the Midwest (Charlotte, Atlanta, Memphis), the strike now provoked local coverage. Apart from the publicity angle, striking the distributors was a brilliant tactical move that provided a greater economic impact per struck job. Bending under pressure, GM eventually agreed to bring all workers at its aftermarket parts depots up to the rate of production workers. The UAW also persuaded GM to eliminate the lower wage tier that parts distribution workers had been placed on in earlier rounds of concessions.[27] The elevation of part distribution jobs represents a tactical victory and discursive coup on the part of labor.[28]

Dealerships

The UAW sent community supporters to dealerships to rally support. Previously, the UAW had been overly cautious, seeking to avoid antagonizing suppliers, most of whom were not owned by the Big Three. In fact, U.S. labor law, under its secondary boycott provisions, heavily restricts workers' ability to picket any site other than their immediate employer.[29] But with three-quarters of Americans supporting the strike and awareness already near 90 percent, the UAW calculated it could afford to spread its reach further, even if it risked alienating a few consumers.

Temps

Like VW, the Big Three rely on supplemental workers to provide ready labor power without the costly training to which permanent employees are contractually entitled. Temps also serve a secondary screening function, filtering out workers who are not cut out for the rigors of assembly line work before the company incurs sunk-cost human capital investment. Apart from their significantly lower starting salary, temporary workers offer substantial cost savings because they do not qualify for salary increases, profit sharing, or pension benefits, and they receive only two to five paid days off annually.[30] Recognizing the importance of temp work to the company's business model, the UAW did not ask to eliminate it entirely but sought to convert long-standing temporary employees to permanent status. Still, workers reacted enthusiastically to the news that Big Three temps with more than ninety days would be converted to permanent status immediately, while future temps would become permanent after nine months.[31] Though pay increases would occur gradually, the commitment to making temps permanent laid a path for eliminating the subclass labor that journalists have described as the third tier.[32]

Back to the Future

Though Fain gets all the attention, his persona is arguably less important than the structural and tactical changes he has overseen. That message would be brushed aside without a mobilized rank and file. The UAW encouraged each factory to make the strike its own, and even plants that continued working through September and October found other ways to participate through practice pickets and solidarity convoys.[33] Fain called on workers at nonstruck plants to refuse voluntary overtime, which is permissible under an expired contract.

Innovation connects the urgent temporality of the day-to-day with the *longue durée* of institutions, drifting between legacy and emergence and back again. Under Fain, the UAW has tried to reinvent itself by harking back to its glory days. Fain invoked the stand-up strike (a play on the sit-down strike, among the most celebrated events in American labor history[34]) not because he wanted to trot out organizational lore for its own sake or to mythologize past struggles in all their messiness and imperfection, but because the past bears on the present. The sit-down strikes exemplified solidarity, as they de-centered and complicated the spatial relations of the workplace. Event time cannot draft blueprints for the future without a furtive glance backward; its temporal movement has a depth that is as sentimental as it is anticipatory.

Union Democracy

Union democracy remains a divisive topic, even among respected labor scholars. Kate Bronfenbrenner et al.'s *Organizing to Win* is an era-defining text that has served as a how-to manual for unions aiming to stem the tide of organized labor's decline and remains essential reading for those invested in new organizing[35]. But for all its merits, the book has precious little to say about the formal characteristics of organization, focusing instead on the details of strategic planning and the tactical efficacy of organizing approaches. The book is essentially agnostic on questions of leadership and rank-and-file power. To be fair, its silence on the contentious subject of union democracy undoubtedly boosted its popularity. For example, *Organizing to Win*'s broad appeal was evident during a fractious rivalry within the SEIU when Sal Roselli's SEIU-West challenged and ultimately split from its parent organization under threat of receivership. The rival groups disagreed on just about everything except their commitment to the organizing model, demonstrating that entrenched bureaucracies could adopt its precepts just as easily as upstart, dissident breakaway movements.

Though union democracy has defenders across the political spectrum, the definition of the term is by no means clear. In Seymour Martin Lipset's seminal text, formal *procedural* democracy—especially leadership turnover and constitutional provisions for contested elections—serves as the primary indicator.[36] But direct elections are necessary but insufficient democratizing instruments. Today, many unions maintain de facto one-party rule despite allowing direct elections. Less commonly, some unions manage contested elections even with a delegate system in place. Absent rival factions with the strength to field viable candidates, direct elections only serve to provide the ruling party with a notarized stamp of approval.

Breaking with Lipset and drawing on a European tradition stressing deliberative democracy, Judith Stepan-Norris and Maurice Zeitlin incorporate additional measures of membership participation.[37] For Stepan-Norris and Zeitlin, democracy has three essential components: procedural (a democratic constitution), compositional (institutionalized opposition), and processual (an active membership). These are not optional criteria but obligatory prerequisites; all three must exist in equal measure for democracy to blossom. But as Stepan-Norris and Zeitlin note, there is significant concurrence among political theorists that favorable constitutional environments create space for contestation, just as contestation gives rise to formal democratic rules. Yet of these three criteria, Stepan-Norris and Zeitlin emphasize rivalry and factionalism, which they see as democracy's defining features.[38]

Union radicals standing in "perpetual opposition" to their leadership are vulnerable to the charge that by positioning members against one another, they weaken the cause of unionism. However, as labor sociologist Judith Stepan-Norris has convincingly suggested, far from weakening unions, internal insurgency strengthens them by forcing leaders to adopt a more militant, democratic posture. Stepan-Norris's work shows that in the 1930s through 1950s, the unions with the most internal factionalism were the ones that negotiated the best contracts.

Even Lipset, though more interested in procedural democracy, anticipated much later work on dissensus and agonistic democratic forms, seeing factionalism as nothing less than "decisive proof of democracy"; internal conflict was the "lifeblood" that might sustain a vibrant organization.[39] Unity is fascism's handmaiden; democracy demands its opposite. According to Norris and Zeitlin, union democracy cannot be reduced to the formal rules for selecting leaders. An organization's democratic character derives not from its bylaws but through a contested power struggle that is at democracy's core.

Under the AC, the UAW exhibited none of the democratic attributes Stepan-Norris celebrates. In its zeal to perpetuate its rule, the AC turned to undemocratic instruments and installed autocratic, highly centralized leadership. By concentrating power in the hands of those positioned to develop works councils and transnational alliances, the UAW ensured that the powerful would remain in power. The AC would seem to confirm what Robert Michels has described as the iron law of oligarchy, or the tendency for organizations to consolidate power over time. In his extended study of German trade unions under social democracy, Robert Michels argues that large organizations have an inherent tendency toward rigidification over time—the iron law of oligarchy.[40] However, reading Michels, one gets the sense that some "natural" tendency toward corruption exists independently of capitalist labor relations or history. For Michels, oligarchy is much more than a risk or tendency; it is "an inevitable current," an "immanent necessity" inherent to the character of mass organization.[41] No amount of grassroots activism can rescue organizations from their predestined fate.

Stepan-Norris and Zeitlin are no strangers to the danger of oligarchy, but they see it, like democracy, as a contingent development dependent on determinate and measurable intraunion variation, not a foregone conclusion. Maintaining a culture of democracy requires (1) the preservation of an active caucus structure independent from officeholders and (2) contested elections to keep leaders and caucuses accountable. If UAWD were to consolidate its power and style itself after its predecessor, the UAW would resume its comfortable antidemocratic pattern, no matter how many paeans to democracy have come out of Fain's mouth. Fortunately, this is not likely to happen anytime soon. Even today, UAWD's grip on power is tenuous. UAWD-supported candidates currently hold only seven International Executive Board seats, compared to six AC members; they have a majority by the barest of margins.[42]

The UAW's commitment to democracy goes beyond its assent to the formal, parliamentary principle of "one member, one vote." While tentative agreements are often presented to the membership as faits accomplis, requiring only a perfunctory sign-off, the UAW, at considerable risk, has encouraged genuine debate over its tentative agreements, even as its contract with GM was nearly rejected at several key plants. In 2014, VW workers found reasons to distrust the organizing team assigned to their plant. Fain's deep commitment to democracy is notable because even defenders of democracy sometimes shy from the messiness that a true democratic culture must entail. To his credit, Fain cultivated a culture of transparency that tolerated internal dissent, at least within limits. He also developed innovative strategies to keep members engaged. For example, the union employed "practice pickets"—simulated mobilizations that prepare workers for their eventual deployment while also signaling the workers' resolve to the company—from the Teamsters.[43]

Whereas previous leaders had presented tentative agreements as *faits accomplis* or used high-pressure tactics to encourage approval, Fain has generally been eager to cultivate a culture of democracy, openly encouraging robust debate and discussion among the ranks despite the messiness entailed. The UAW's "historic victory" was achieved by only 54 percent at GM, and the proposed contract was defeated at some of the largest plants, leading some to conclude that Fain was more bark than bite. Nevertheless, the contract's slim margin of victory was perhaps the best indication that a democratic culture had pervaded the organization at all levels, unrestrained by the magnetic pull of organizational loyalty that would have workers rally behind

Fain's cult of personality. Indeed, large margins can mask a sense of resignation or abstentionism. Still, there are limits to Fain's democratic principles. When the membership rejected an agreement that fell short of expectations at Mack Trucks, Fain initially responded graciously, saying that he was "inspired to see UAW members at Mack Trucks holding out for a better deal."[44] But after holding out six weeks with little progress, the UAW returned the same agreement to the membership, evidently having concluded there was nothing more to be gained, and used strong-arm tactics to encourage approval.

Though UAWD's future remains unwritten, at least through 2024, it continued to function as a dissident caucus with an active base on the shop floor even though it now has friends in high places. It has proved capable of independent action above and beyond what its leaders might condone. For example, UAWD-aligned workers at Stellantis delivered a petition to management protesting its failure to convert temps to permanent status—all without official sanction from International Headquarters. As is always the case, the survival of democracy in the UAW will depend not on formal rules or on the leadership's willingness to tolerate dissent but on the perseverance of a mobilized rank and file that unrelentingly demands accountability, no matter who sits atop Solidarity House.

Contention

Key to UAWD's effectiveness is its understanding of power and power relations. Though its representative now holds power, UAWD has preserved a semiautonomous existence as a caucus to prepare for future elections and ensure Fain and company remain accountable. UAWD has used its leverage to advocate, occasionally publicly disagreeing with Fain when his position was compromised. Thus, even once elevated to a leadership position, Fain has continued to face internal pressure from UAWD, the rank and file, and the federal monitor overseeing his organization. Competitive environments under constant threat from insurgents provide the substrate where democracy can gestate. Even the hard-charging militant becomes a calcified fossil without disruption from below. Put differently, Fain's militancy is attributable not to his personal style or ideological commitments but to the pressure-cooker environment that produced him and continues to make demands on his administration. Fain's ability to hold on to power will depend in part on his ability to appease the militant wing of UAWD, not to mention his effectiveness at triangulating the expectations of an increasingly vocal rank and file.

Concurring with Lipset, Stepan-Norris and Zeitlin argue that democracy thrives when leaders embrace a "diffuse political consciousness" aimed at the general uplift of the working class instead of narrow or self-interested goals. This "transcendent conception of the unions' mission" is tailored to provoke controversy and trigger dissent—itself a democratizing force.[45] It matters little whether the union is willing or capable of backing its political consciousness with action as long as its political positions are sufficiently polarizing to produce democratizing effects. On this count, Fain also meets the criteria, borrowing heavily from the Bernie Sanders playbook and embracing a soft democratic socialism with populist overtones. Fain's call for a ceasefire in Gaza stopped well short of mobilizing his members to build weaponry for Israel to enforce such a ceasefire, but nonetheless fomented controversy with democratizing consequences. Such resolutions are risky for leadership as they enrage opponents while providing scant short-term benefits, but by the Stepan-Norris and Zeitlin standard, their democratizing potential is nearly unparalleled.

Rank-and-File Strategy

As unions grew more bureaucratic and conservative in the 1960s and 1970s, rival caucuses formed in many public and private sector unions, challenging the leadership from within. The best-known caucuses were in the UAW and the Teamsters, but groups also formed in the United Steelworkers, the American Federation of Teachers, Communications Workers of America, and many other unions. These movements challenged the racism and sexism of union bureaucracies, fought for better contracts, and organized militant strikes and other job actions, often without permission from union leaders. But in nearly every case, running for office was the centerpiece of the strategy.

By its own standards, the rank-and-file movement failed. Dissidents took over some important unions but rarely held them for long, at times preferring their freedom to the responsibility and respectability politics that came with managing an organization. By the end of the 1970s, the rank-and-file movement had fizzled out in all but a few unions. In only one union did left dissidents succeed in taking over the international organization: the United Mine Workers. (Reformers would eventually take over the Teamsters, but this occurred much later and in a different context.) There was also an attitudinal shift: "a new disregard for the legal and disciplined practices of modern trade unionism, and a deep suspicion of hierarchy manifested through a decentralization of bargaining."[46] After years of domination, the rank and file were in a state of open rebellion against their unions, and leaders often had trouble keeping up with the accelerating pace of demands. These transformations had important, if fleeting, consequences, but in the eyes of most reformers, they were secondary to the real political project: taking power.

Barbara Smith observes in her now-classic history of reform movements in the United Mine Workers of America that few rank-and-file caucuses have survived this transition from the streets to the boardroom. Upon winning top office in the 1970s, Miners for Democracy (MFD) collapsed itself into the organization it had taken over, eventually becoming indistinguishable from the very bureaucratic apparatus it now controlled. In Smith's view, this decision was fateful: "It left the new administration without a coherent rank-and-file base, and it left the rank-and-file without an organized vehicle to hold their new leaders accountable."[47] Absent an independent locus of power, the reform impulse that had propelled MFD into power was quickly subsumed beneath the day-to-day responsibility of governance. MFD found itself with no recourse as its leadership slate became increasingly isolated and ineffective. Reform caucuses are naturally loyal to leaders that emerge from their ranks, but their persistence beyond election night is essential to the long-term success of their agenda.

Like countless insurgent groups in the union movement, MFD cannibalized itself when transitioning from opposition to administration. Without an independent power base, MFD had no mechanism to ensure accountability, safeguard its victory, and avoid backsliding or counterreform by the old guard.[48] Even the *Pittsburgh Post-Gazette* opined:

> One could argue . . . that now that MFD is in the saddle, it hardly can be a rebel outfit against itself. But this actually points to a continuing need in the labor union for opposition.[49]

In a recently published essay collection on labor militancy in this era, former movement activists freely acknowledge that the movement's singular focus on attaining office contributed to its ultimate failure:

> Newly elected officers remained subject to the same constraints as their predecessors, facing pressure from employers without the permanent self-organization of the rank and file that would provide strength and discipline to individual leaders.[50]

Ironically, the origins of the rank-and-file reform movement can be traced to neo-Trotskyite sects, such as the Johnson-Forest Tendency, which promoted "self-mobilization of the working class" independent of traditional party politics and union leadership. Echoing this influence, a minor current within the reform movement opposed the focus on electoralism. But for other elements within the sectarian Left, taking over unions was understood as a prerequisite for seizing the state apparatus and implementing socialism. Over time, the pro-electoralist position came to dominate the rank-and-file scene, and reformers soon grew obsessed with and infatuated by power.

The American union movement has often been torn between its dual role as a seller of labor power and manager of organizational affairs. The union movement lacks a model for a non-despotic form of power. Even the book *Democracy Is Power*, written by allies of New Directions and intended as a handbook for rank-and-file rebels, defines democracy principally as a question of ousting entrenched bureaucrats and replacing them with reformers.[51] Nevertheless, the evidence favoring the rank-and-file strategy is abundant and clear. Thaddeus Russell argued convincingly that the Teamsters' militancy under Jimmy Hoffa in the 1960s was due to a rank-and-file activist who was willing to engage in militant activity but had no designs on power. In an article for New Labor Forum about the rank-and-file caucus within the Teamsters that later transformed itself into a ruling party, Thaddeus Russell writes:

> For union members who want higher wages, less work, and better conditions, the historical evidence suggests that a strategy of perpetual opposition to the union leadership, regardless of its background or intentions, might be more effective than attempting to become the union leadership. According to this analysis, union democracy is best viewed as an instrument of leverage with which to force the bureaucracy to perform, not as a means to install a new bureaucracy.[52]

While one cannot expect a deeply subsumed organization to extricate itself from the throes of domination, there are strategies available to rank-and-file activists who seek to dismantle domination from within. These strategies together constitute a renegade subjectivity—a mode of activism that requires immersion inside an organization but not allegiance to that organization. That is, renegade subjectivity operates in a conceptual space that is both *inside* and *against*. Renegades are disloyal to their organization's leaders but have no designs on power themselves. Their force is derived precisely from their embeddedness. The rank-and-file movement is a living mode of engagement that is fragmented, recombinant, and constantly reopening to the outside.

As noted above, winning office has been the downfall of most rank-and-file movements. Early rank-and-file movements often ran candidates for office, but they rarely won. Fain and company now find themselves in a paradoxical position with respect to power. There is perhaps an inherent tension between the antisystemic allure of rank and filism and the structural imperatives of governance. Fain is now simultaneously the product of a fringe rank-and-file faction and the leader of an organization with more members than the city of Cincinnati. The compromises necessitated in the transition from belligerent activism to the somber task of managing an organization bound by the strictures of an elaborate legal apparatus have destroyed many a rank-

and-file movement. Meanwhile, the seductions of power entice even hardened militants to abandon their ideals at the door of the union hall. Navigating these competing pressures is an inherent contradiction.

Conclusion

This chapter has shown that the UAW had to break with the past to reinvent itself. Though Ganz uses the terms *creativity* and *innovation* more or less interchangeably, following Schumpeter, I have split innovation into an affirmative impulse (creativity) and a negative impulse (destruction), themselves dialectically linked. Through the dark arts of "creative destruction," UAWD launched an autoimmune defensive attack on the organizational body itself so that it might someday heal.

As is typical, the impetus for innovation came from the rank and file, not from "enlightened" leaders acting on their own accord. Lacking such internal ferment, unions become susceptible to oligarchic tendencies, as Ganz's case study highlighted. Though his model is built on the heralded United Farm Workers of the 1960s and 1970s, when grape farmers were a cause célèbre for New Left radicals, he devotes a few pages to the organization's decline.[53] The absence of formal democracy within the United Farm Workers lay behind its devolution to a cult of personality in the 1980s and its near disintegration in the years that followed Cesar Chavez's death. Innovation (here, "heuristic facility") dwindled as the organization became set in its ways, foretelling a slow decline that dragged the organization toward stasis even though learning acumen remained strong. Innovation spurs organizational renewal by deposing old, ineffective forms and replacing them with new ones.

Relatedly, this chapter has taken into account the dialectic between effective leadership and mobilized workers. Ruth Milkman has argued that "successful organizing must be comprehensive in scope, combining and synthesizing top-down and bottom-up approaches."[54] The pre-2023 UAW possessed neither bottom-up participation nor competent leadership and ended up with the worst of both worlds. In a somewhat different register, Nelson Lichtenstein's study of Walter Reuther often centers on the fraught and codependent relationship between the UAW's leadership and its insurgent rank and file. Union officials derive their bargaining power from the possibility of disruption, Lichtenstein suggests, even as the terms of the labor regime require them to control that threat at the behest of management. In the long term, power is thus intertwined with and inseparable from its opposite, control.[55]

Finally, this chapter pulled back the analytical lens to revisit UAW from an organizational studies perspective, stressing how the organizational form aligns with its content. The dismantling of the AC and its replacement by a new generation of reformers surely infused the organization with energy. But the UAW's rank-and-file rebellion set in motion a broader set of changes—crosscutting the strategic capacities already highlighted—while innovatively proposing a new relational basis for solidarity. The subsequent chapter traces the implications of these transformations in Chattanooga.

The UAW's rank-and-file rebellion was transformative, not simply because it installed new leadership but because it signaled real change beyond surface-level remedies and cut to the heart of the UAW's organizational identity. Strategic choices that reaped rewards in 2024 (a tough stance against the Big Three, the inclusive and representative organizing committee, and a renewed commitment to transparency) reflected broader structural changes that altered the organization's internal culture and rendered old habits obsolete. Given the structural character of the UAW's transformation, it is not clear that the UAW of 2014—bound by its organizational limitations—would have been capable of the same outcome. The 2023 stand-up strike allowed for a fleeting moment of collective amnesia, capturing the imagination of the newly mobilized rank and file and diverting attention from the recent history of defeat. The opposition once again sought to dredge up the organization's earlier struggles, but the UAW now had a ready-made rejoinder.

7

SOLIDARITY REINVENTED?

How the UAW Won in 2024

Despite improving its strategic capacity on three of four counts, the UAW lost the 2019 vote by nearly the same margin as in 2014. Examining the 2019 vote led us to a discussion of organizational failure and learning. The UAW's shortcomings in 2019 were deep-seated, structural, and endemic, characteristics that reduced the possibility that they could be remedied through iterative learning processes alone. Organizations facing such systemic problems must innovate to escape from path dependency.

This chapter extends our strategic capacity model to the 2024 election. On the management side, obstacles to unionization decreased again in 2024, improving the union's prospects dramatically. Whether this alone would have been sufficient to deliver a win to the UAW can only be a matter of speculation. Meanwhile, the union maintained good access to knowledge and strong motivation, much as it had in 2019. However, to seal the deal, the UAW could not settle for boosting its learning quotient alone. Apart from enhancing its heuristic facilities, it would have to innovate to liberate its future from the shadow of the past. The UAW's ability to innovate allowed it to break free from the cycle of past defeats and address the systemic problems that had plagued its previous organizing drives.

This chapter explains the 2024 victory by highlighting the shifts that improved strategic capacity beyond the more limited gains of 2019. The chapter's core systematically evaluates the 2024 win as it compares with previous losses. The highlights of 2023 and 2024—the democratic revolt at the 2023 UAW convention, Fain's unlikely rise to power, the 2023 stand-up strike, and the renewed push to organize transplants in spring 2024—can only be examined in aggregate. I treat these transformative events as a continuous progression, not only because they occurred on an accelerated timeline (just over a year) but also because they were thematically interrelated.

Unlike the chapters on the 2014 and 2019 elections, this chapter's strategic capacity analysis begins with innovation and then covers motivation before proceeding to knowledge and learning. This foregrounds innovation, which proved the secret ingredient necessary for victory: it was absent in 2019 and misdirected in 2014. The strategic capacity measures are presented "out of order" both for reasons of chronology and because innovative practices prefigure knowledge, learning, and motivation. All four dimensions of strategic capacity became linked through synergistic feedback loops and reciprocal coeffects.

Obstacles

In 2014 and 2019, management's overt and covert union busting, a well-funded in-plant opposition campaign, and elected leaders' political interference made the UAW's task exponentially harder. Overt anti-unionism continued apace in 2019. By 2024, all three obstacles had diminished somewhat, and workers were better prepared and inoculated against threats.

Union Suppression

As during previous elections, the UAW identified multiple unfair labor practices in the run-up to the vote. Workers documented management interference with flyering, policing of conversations, at least one captive audience meeting, and other misconduct.[1] This conduct did not approach the frequency or severity of 2014, but it serves as evidence that the company had—for a third time—reneged on its neutrality pledge, despite its official protestations to the contrary. Workers had become appropriately skeptical of corporate communications and no longer placed much stock in these official announcements. Still, there can be no doubt that the intensity and frequency of union suppression lessened in 2024. Even in its public statements, management was remarkably restrained in its criticism of the UAW.

Promises and Improvements

Change has been a constant at VW since the plant opened. Long-term employees have witnessed the frequent turnover of top-level managers, repeated adjustments to schedules and scheduling practices, and fluctuations in staffing levels. But as in 2014 and 2019, VW pushed through changes on the eve of the union vote.

The biggest change was a direct consequence of the stand-up strike. Following a pattern Fain referred to as the "UAW bump,"[2] VW increased pay in direct response to the Big Three contracts. Just after the deals were ratified, wages at VW were boosted by 11 percent, rising to $23.42 hourly, matching the 11 percent immediate increase Detroit workers received after ratification, though still falling short of Detroit's new $35.26 base rate since pay at VW was lower to begin with. Though some saw VW's 11 percent payout as little more than the corporate equivalent of hush money, the raise was substantial. VW workers still made less than their Detroit counterparts, but their pay was already competitive for the area.[3] Moreover, on top of the raise, management compressed the wage scale, allowing starting employees to reach the maximum rate more quickly. The "UAW bump" confirmed what union advocates have long insisted: organized labor drives up industry wages and affects unionized and nonunionized workers alike.[4]

During previous campaigns, though wages had long been among the most significant factors separating foreign automakers from unionized American firms, in its efforts to organize transplants, the UAW had assiduously avoided the issue of pay, surmising that injuries, overtime, scheduling, and other ancillary issues were of more concern.[5] The vast pay gap between transplant workers and those with Big Three contracts was downplayed by spokespersons and largely absent from the UAW's formal messaging, even as the wage differential increased over time. Workers with prior union experience were well aware that unions had negotiated regional pay scales, but their concerns went unheeded.

After the 2024 election announcement, VW also promised to modify a PTO policy that had previously charged workers for seasonal shutdowns. Though the change had been a core union demand, it was relatively inexpensive for the company since the plant had already been fully retrofitted to produce EVs. While this change was not insignificant, workers had become desensitized to the preelection pattern of incremental improvements. They realized that without a contract, management had the latitude to modify its policies at its discretion, implementing or eliminating popular programs to suit its needs. This time around, since workers anticipated a major victory on par with transformative wins at the Big Three, few were content to settle for a slight marginal improvement.

Political Interference

In 2024, the local political class maintained its opposition but to less effect. Lee, now serving a second tour of duty as governor, returned to the plant for a Hail Mary press conference but was not permitted to enter the gates and had no choice but to deliver his remarks from a semipublic plaza facing the employee parking lot. In a sign that the UAW had already shifted the terrain, Lee partnered with three other southeastern governors to issue a joint statement. The UAW had persuasively argued that the dozen or so foreign-owned assembly plants were not stand-alone entities but interconnected nodes in a global production system. In responding jointly, the governors had already conceded this point.

Further, the opposition's messaging had grown increasingly desperate. The governors claimed, "Every single time a foreign automaker plant has been unionized, not one of those plants remains in operation."[6] Though technically accurate, the statement failed to acknowledge that these closed plants had been cooperative joint ventures with American firms, often operating as quasi subsidiaries of the Big Three, as in the case of Saturn. Moreover, when politicians dusted off their talking points from 2014, their message of union-driven deindustrialization now hit differently. In evoking "Detroit," politicians only reminded workers of the landmark Big Three contracts.

Compared to the full-scale media war waged in prior elections, the 2024 effort was paltry. Amid a pro-labor climate, opponents may have feared inadvertently boosting support with a heavy-handed reaction. To be sure, some of the same imagery reappeared (the billboards, the lawn signs), but it seemed half-hearted, pro forma—almost resigned to defeat. One possible explanation: for the first time in forty years, Tennessee was playing host to a unionized automaker at a greenfield construction site. With $6 billion invested in Ford's Blue Oval City in West Tennessee, the state may have feared angering Ford by taking a hard-line approach to the UAW. Moreover, the opposition was running out of talking points. It reemerged in diminished form, a broken record mechanically repeating its greatest hits from 2014, less a revival act than a tribute band. Their website's very title spoke volumes: *Still* No to UAW.[7]

Inside the plant, the opposition was even more muted. What little resistance arose was easily overpowered by the momentum of the organizing campaign. The UAW made a real effort to engage opponents this time—another break from the past. As workers recounted, even previously hostile opposition leaders became ripe targets for organizing, allowing the union to attract new adherents instead of shoring up its base. Some former antis even supported the union after watching the success of the stand-up strike: "Watching the Big Three is one of the things that changed their minds."[8] Even when opponents mobilized, their message was unconvincing. There was a different vibe by 2024, tilting even ardent opponents toward a more pro-union stance. As one union supporter noted:

> It's a lot different this time. Third time's a charm. The people who were on the fence . . . now they can get on board. Now, they say, "I want a union. Tell me what I need to do."[9]

Workers also reacted to political interference with the shrewdness of veteran organizers. Another commented: "How dare a politician who's never stepped foot in this plant who probably cannot do the jobs that we do every day."[10]

Disinvestment

The threat of disinvestment had loomed large over previous elections in 2014 and 2019. In both instances, politicians intimated that a yes vote would force the company to cancel planned expansions or possibly uproot the entire plant in search of more submissive workers. Billboards depicting abandoned factories and implying that Chattanooga could expect a similar fate if it unionized reinforced this message. The calculus never quite added up; given the enormous sunk costs associated with a greenfield auto plant and the specialized layout designed for the company's proprietary modular assembly system, the possibility of relocation had always been slim. Nonetheless, the company's popularity and the factory's outsize impact on the region made these threats resonate.

In 2024, the specter of disinvestment reared its head once again, but under dramatically different circumstances. Previously, Chattanooga had ranked among the least productive plants in the country, posting inconsistent profit margins and rarely meeting its output quota. This changed in 2022 when the plant was awarded VW's first (and at the time, only) EV targeted toward the North American market, the ID.4. With this vehicle, VW would become the first international automaker to qualify for a $7,500 federal tax credit in 2024 and further seal the company's commitment to the region. As the sole producer of VW's flagship EV, Chattanooga had become essential to the company's ambitious goal of achieving 55 percent electric-powered vehicles as a portion of total sales.[11] VW soon reported "overwhelming demand" for ID.4 and announced plans to double its sales—not typical news for the beleaguered plant.[12] With the ID.4 representing 11.5 percent of its total U.S. sales, VW could not afford to eliminate the model. Doing so would also have prevented the company from meeting the upcoming 2027 tailpipe emissions standards.[13]

Moreover, by 2024, VW had established a strong presence in the region, making the prospect of relocation even less credible. Whereas the company had initially sourced many of its parts from overseas and Mexico, the maturation of the southern auto cluster, combined with a push to reduce shipping costs and minimize risk, led to the shoring up of its supply chain network by 2024, with 90 percent of suppliers located in the Chattanooga metropolitan statistical area (which extends over state lines into northern Georgia and Alabama). The tightly linked supplier network it had worked hard to cultivate would be difficult to abandon. A dense network of EV-related suppliers, many supported by taxpayer dollars, had cropped up in the area. Because of strategic coupling between VW and its host city, now backed by major investments and supported by numerous stakeholders, capital flight was no longer a real threat—if it ever had been.

Since 2014, Chattanooga had transitioned from a middling, underperforming, low-capacity plant making increasingly unpopular midmarket sedans to a core node in the global production network building heavily incentivized EVs essential to the company's market strategy. For all practical purposes, the plant had become irreplaceable. Since 2014, it had doubled its workforce, added a new production line, invested $800 million in new capital, and converted to EVs. In 2014, the possibility that VW might divest from Chattanooga seemed remote but plausible. By 2024, it would not take an accountant to realize the notion that VW might cut its losses on $3.9 billion in fixed capital was laughable. As one worker said:

> They are getting ready to bring out new models. They will not do all this preparation just to pick up and leave. That is enough to tell me that VW is not going anywhere, union or not.[14]

This was a stark difference from 2014, when the plant was struggling and ramping down production, and 2019, when it was just beginning to meet its targets. In the long view, employment and productivity steadily increased over the decade.[15] With the Chattanooga plant now the key link in VW's "green transition," job security was no longer an issue, employment increased manifold, and total production set records in both 2023 and 2024.[16]

As we have seen twice before, elections tended to coincide with conveniently timed production-related announcements that lent credence to the threat of disinvestment. This time, the stars seemed to align for the UAW. There could be little room for misinterpretation when management described the ID.4 in effusively optimistic terms, with one executive saying, "Frankly, we're just getting started. It's our mission to lead the industry in a new era."[17] Shortly before the 2024 election, the company announced it would move to a third shift, hire one thousand additional employees, and aim to boost its output by 25 percent.[18] With VW jobs now secure and positions being added at an unprecedented scale, the governor's relocation threat no longer stood up to scrutiny: "Unionization would certainly put our states' jobs in jeopardy . . . in this year already, all of the UAW automakers have announced layoffs."[19] This point might have resonated had it not been for VW's binding commitment to EVs, which eliminated the threat of disinvestment in the short and medium term.

General economic conditions also favored the UAW. Research on the impact of profitability on unionization is complex, but transformative social movements tend to emerge during times of abundance, while defensive movements are more likely to occur in periods of scarcity.[20] Moreover, tight labor markets have been associated with emboldened workers and higher unionization rates due to the lower availability of replacement workers.[21] The pandemic-era labor market favored unionization—a pattern that lasted at least into early 2024.

Innovation

Innovation is not divorced from history but requires a willingness to trade the security of familiar routines for the wager of speculative futures; it *transcends* the past rather than relitigating past defeats. Seen through the lens of innovation, the 2019 election drew the wrong lessons from 2014, overcompensating for its tactical disadvantage by allowing the opposition to define its campaign. Still reeling from the 2014 defeat, it retreated into a defensive posture, circling the wagons to defend its ranks against an all-powerful enemy. By 2024, the collective trauma of 2014 had receded, and the UAW was ready to move forward, not backward.

Scholars of organizational creativity often observe that innovation comes from unexpected places.[22] In the case of the UAW, the innovative spark came initially not from the Organizing Department, which is formally tasked with expanding membership rolls, but from its National Collective Bargaining unit, which oversees expired contracts. Indeed, as with all unions, functional units are organizationally distinct. But prospective members are not likely to be much interested in the UAW's organizational charts. Instead, they draw commonsensical connections between the UAW's past performance and future results. By explicitly linking strong contracts with new organizing, Fain broke sharply with his predecessors, who had tended to present "new organizing" at transplants as distinct and siloed off from the UAW's core obligation to represent its existing members.

In 2014 and 2019, the "problem" of the southern transplants was firewalled within the Organizing Department. The 2023 strikes showed that efforts to grow the organization cannot be divorced from self-sustaining operating expenses (not least because the organization cannot survive without increasing its membership rolls as automation and outsourcing continue to chip away at its Big Three base).[23] Put differently, bargaining representatives knew their success at the negotiating table was crucial in advertising the union's accomplishments to would-be members, while givebacks and concessionary contracts weakened the union's sales pitch. From an organizational studies perspective, under the AC, siloed thinking, noncollaborative work processes, and inertia had confined the organizing-related responsibilities to the department assigned to these tasks. Under UAWD reforms, the necessity of organizing permeated all levels of the organization and became an inescapable priority, no longer the exclusive province of designated organizers.

As noted in Chapter 6, the UAW's 2024 election was immediately preceded by the ouster of top leadership by an insurgent rank-and-file caucus, sweeping democratic reforms, and a momentous stand-up strike. Though these events were exogenous to Chattanooga, they had a tremendous impact on the organizing campaign. The connection between the UAW's reinvention and the organizing campaign at VW is less a question of direct transfer of tactics than of ambient spillover. Striking for recognition at VW, while technically feasible given the multiple unfair labor practice charges, was never seriously considered. However, the heightening of contradictions and escalation of tactics contributed to an expanded sense of what might be possible. More than an atmospheric vibes shift, the stand-up strike signaled to prospective members at VW that the UAW was an *innovative* organization, no longer content to sink into familiar patterns and quite unlike the moribund union they had twice voted against. The UAW's transformation was not immediate, nor was it total, but it still represented a real break with the past, sufficiently meaningful that even former opponents could not help but give it a second look.

Legal restrictions narrowly circumscribe acceptable conduct for organizing workers in the United States, limiting the possibilities for innovation during a conventional organizing drive. For this reason, many unions confine themselves to the type of iterative learning that Ganz recommends, seeking not to rewrite the playbook but to make slight alterations well within the scope of widely accepted "best practices." Certainly, the UAW continued to rely on the methodical deployment of tried-and-true tactics, accompanied by meticulous recordkeeping that would be familiar to anyone versed in labor organizing. In these senses, the UAW drew on a back-to-basics approach rooted in the unglamorous but time-honored task of worker-to-worker organizing that had been so glaringly absent in 2014.

Still, in at least one important respect, the UAW ran an innovative campaign that shattered expectations and demonstrated real tactical agility despite the risks of testing an unproven strategy in a make-or-break situation. Just as it had changed its approach on the fly to calibrate pressure on particular companies during its stand-up strike, the UAW departed from its habit of one-off campaigns and presented VW as the opening salvo in a broader campaign against transplants. Even as the vote neared in Chattanooga, the UAW was already preparing for another vote at Mercedes, some four hundred miles away, and fielding requests from workers at six other plants by its own accounting. VW workers saw themselves as the connective joint between the Big Three contracts and the remainder of the transplants. They were rightful heirs to the Big Three contracts and the forward line of an extended campaign against the remaining transplants and battery makers. The one-plant-at-a-time approach had meant ignoring interconnections between global companies. The new sectoral focus was a crucial reframing of the campaign that repositioned Chattanooga against a different backdrop.

Industrial psychology has long observed that laser focus on a task ironically leads to less innovative outcomes, and strategic management scholars have suggested that the same applies at an organizational level. As long as the UAW retained its exclusive focus on organizing southern transplants, it could, at best, tweak a narrowly prescribed and bounded set of tactical approaches. As Ganz writes, "For a creative response to be produced, it is often necessary to step away temporarily from the perceived goal to direct attention toward seemingly incidental aspects of the task and the environment."[24] Once it widened the aperture of its strategizing sessions, the UAW realized that the solution was hiding in plain sight at the Spring Hill, Tennessee, GM assembly plant and other Big Three plants. Only by (temporarily) diverting its attention to the Big Three could the UAW pinpoint its errors and develop a new, more expansive tactical repertoire.

As noted in Chapter 6, in both 2014 and 2019, with the organization's calcified structure, antidemocratic governance, and tarnished leadership, it remained stuck in a recursive feedback loop, unable to achieve the velocity that might allow it to escape past defeats. Casting off the shadow of the past required that the UAW turn the reflexive mirror on itself and reconstitute itself from within to open new strategic possibilities.

Motivation

The wave of insurgency and contention that gripped the UAW in 2023 dramatically affected the 2024 election, both through its direct impact and by the fresh life it breathed into an organization many had dismissed as all but moribund. The organization's internal life was reinvigorated in ways that will reverberate well beyond the current contract cycle.

Connections

Ganz notes that nothing is more likely to boost motivation than a string of victories, while a slump produces the opposite effect: "Success augments motivation."[25] Many observers have pointed out that VW drew inspiration from the Big Three strikes. But if the transition between the Big Three victory and a renewed push at VW appeared seamless or automatic to outsiders, it was actually the result of a deliberate strategy. Even before the ballots were fully counted at the Big Three, the UAW had boots on the ground in Chattanooga, testing the waters and working to reform the organizing committee, which had devolved through natural attrition since 2019.

Before the dust had settled on the Big Three contracts, Fain was already announcing "Stand-Up 2.0," his campaign to organize transplants. Immediately after tentative agreements were ratified, the union announced it would parlay its gains into a comprehensive strategy targeting all unorganized firms, foreign and domestic, EV and internal combustion engine (ICE). However, rather than take on employers one at a time—as it had previously—the union declared its intention to organize all transplants simultaneously.[26]

In practice, this meant that campaigns would be staggered to maximize organizational resources. Still, the UAW signaled it had abandoned its piecemeal approach in favor of a simultaneous push across multiple sites. Backing its talk with action, the UAW organized caravans and Zoom meetings for workers throughout the Southeast to share ideas and stories. Just as the UAW had caught the Big Three off guard by dispensing with the one-at-a-time pattern bargaining strategy in favor of a calibrated rollout of strikes across all three companies, it would now run near-simultaneous campaigns at thirteen transplants and EVs, holding elections on an unannounced and unpredictable schedule. It hoped it could generate similar returns by keeping the transplants guessing, just as it had done with the Big Three.

Indeed, the UAW seemed to negotiate its Big Three contract with one eye gazing across the table at the company bargaining chiefs and the other cast southward toward its future enlistees. Its top priorities included cost of living adjustments, which resonated in the low-wage South, and an appeal to young workers, who were likely to stand at the receiving end of the two-tier wage scale. Again, Fain explicitly drew these connections, saying, "We want to use the Big Three contracts as our framework [for transplants]."[27]

The UAW also suggested it would synchronize contract expiration dates to align with the Big Three in 2028, potentially paving the way for a general strike. Previously, campaigns at transplants had often occurred in off-cycle years, usually at least twelve months removed from Big Three contracts. But Fain connected VW with the Big Three both rhetorically and temporally, showing that "unionism in one company" was a doomed prospect from the outset.

Workers contemplating joining a union do not conduct a simple cost-benefit analysis. However, they make a general comparative evaluation, observing pay, working conditions, and benefits in unionized industries to gauge their prospects under a union. In this sense, the Big Three served as the primary reference group for VW workers, putting forward the closest approximation of a unionized future. Fain made this connection explicit. Referencing the wage bump at Toyota, he extended an open invitation to transplant workers:

You are welcome to join our Stand-Up movement. If this is what Toyota gives you when the Big Three stand up and fight, imagine what you could accomplish if you join the UAW and stand up and fight for your-selves.[28]

The strike had the added benefit of proximity to organizing targets. By striking distribution centers, the UAW broadened its reach beyond the final assembly plants, establishing a beachhead in the South, which had only a handful of assembly plants but numerous distribution centers. For example, the stand-up strike extended to a GM distributor just outside Atlanta, only about two hours from Chattanooga.

With the stand-up strike concluded, rather than foreign automakers setting the floor for Detroit and undercutting Detroit's gains, Detroit now set a ceiling for foreign automakers. The strike increased the marginal difference between the Big Three and transplants, incentivizing foreign companies to catch up to remain competitive. Immediately after the strike, other companies rushed to match the Big Three contracts, demonstrating the union's power. VW workers saw their fate as linked to the other southern factories but also understood the central importance of their struggle as "one of the first dominoes."[29] A win at VW would set a new precedent, just as repeated losses had reinforced the old one.

Ripple Effects

As noted in the "Promises and Improvements" section, the strike had an immediate direct impact in Chattanooga in the form of a substantial raise. For the last forty years, the UAW's fruitless efforts to organize so-called transplants were a defensive rearguard measure aimed at preventing foreign automakers from undercutting union wages. The claim that low wages in the South undercut its union contracts and drove down wages everywhere has a solid economic basis and has long stood at the center of the UAW's push to organize transplants.[30]

With the Big Three contracts concluded, the UAW was no longer playing defense, protecting its members from the unorganized South. Instead, the Big Three had now set a new, higher wage standard, which transplants quickly sought to match. With Big Three wages at historically high levels and foreign automakers racing to catch up, the UAW had gone on the offensive, eliminating the regional wage differential and setting a new high standard across the industry. Thus, rather than a race to the bottom, the contracts set off a race to the top, with transplants newly incentivized to compete with Detroit's new rates. Region 8 director Tim Smith put it succinctly, noting that during previous elections, "the union didn't have anything on its plate for workers to look forward to. With what they accomplished with the Big Three, we feel they can accomplish that in the South."[31] A worker echoed this sentiment, noting:

We couldn't point to a time in their lives when they were watching the news and saw, "Oh my God, look what they did. That's amazing. We can do that."[32]

Organizers normally avoid pitching unionization in purely instrumental terms, but the too-simple calculus of cost-benefit analysis worked in the UAW's favor this time. A worker at Mercedes in Alabama observed:

In the past, when we ran campaigns, we would tell people we have to make a marginal argument that the Big Three is getting a little more money; they got pensions and a little more benefits. When you put it on paper, it was a marginal argument. . . . Now the difference has got so much greater since the Big Three has increased its formula on profit-sharing, and got some really nice raises.[33]

The strike was at the forefront of many VW workers' minds. It had dominated the national news, headlining major outlets throughout October, so it would have been a popular watercooler topic in almost any setting. But at VW, the strike's proximity and obvious relevance for other autoworkers magnified its importance. One worker said:

We pay close attention to it, and then seeing that you have people like us, that make cars every day actually win outstanding contracts. It made you think.[34]

Workers reported that the strike was a constant topic of conversation and speculation throughout the fall: "People here were rooting for [the strikers]. It showed what you can achieve."[35] On at least two occasions, the first shift even recalled surreptitiously watching Fain's live stream on cell phones. The more innovative demands of the 2023 strike, including the thirty-five-hour work week, were especially impactful in Chattanooga. Workers did not have to make these connections themselves. Fain explicitly tied his strategy at VW to the Big Three, saying:

The Big Three say they can't compete with non-union companies that keep wages and benefits low, so we're going help them out with that problem. We're going to raise the standard across the country, instead of lowering it. Once all the workers everywhere make a fair wage with decent benefits and real job security, companies won't be able to pit autoworker against autoworker to compete with one another for profit.[36]

In 2014, lacking a vibrant union movement to emulate at home, workers looked overseas for inspiration. By 2024, unlike previous campaigns, where VW union organizers felt isolated, they considered themselves part of a nationwide movement. As one Chattanooga community organizer noted:

> It's happening all along the country . . . and that organizing energy is being immediately translated towards major organizing efforts all across the South.[37]

Spring Hill

GM's Spring Hill plant just outside Nashville was of special importance. Spring Hill is today among the company's largest production facilities, having previously built the now-discontinued Saturn line, which ended production just seventeen months before the Chattanooga plant opened. VW actively recruited laid-off Saturn workers, especially skilled technicians, who were sparse in the Chattanooga region. The purpose-built company town is an iconic symbol of the state's burgeoning auto industry and an important reference point for autoworkers in Chattanooga.[38]

Union communications from 2014 stressed Chattanooga's absence from the company's works council network and drew unfavorable comparisons between VW's American workforce and its unionized plants abroad. In contrast, documents from 2024 shifted the reference group and the basis for comparison. For example, a released video mentioned the UAW's unionized plants "up the road" in Spring Hill and Louisville, Kentucky, but made no mention of Germany. In a hyperbolic flourish, one VW worker even referred to Spring Hill as a "shining beacon" lighting the path for VW workers.[39] Likewise, the union's main propaganda instrument, a glossy twelve-page pamphlet, included no mention of the works council but copious references to Spring Hill workers now enjoying the bounty of the 2023 strike and having emerged as standard-bearers for the auto industry.[40] Pushing back against perceptions of the UAW as a third party with no real connection to the South, the broadsheet noted that Spring Hill's GM complex employed seven thousand UAW members and explicitly compared Spring Hill and Chattanooga on issues most important to VW workers. Another circular titled *Improving Work-Life Balance at VW* noted that GM's PTO policy compared favorably to VW's:

> At GM's Spring Hill, TN facility and other GM facilities, the company can only ask workers to use PTO for one week of shutdown per year. After that, they get "SUB pay" on top of state unemployment benefits, which together average 95% of weekly after-tax pay.[41]

VW workers no longer had to look across an ocean for inspiration; the best argument for the UAW was just "up the road" in Tennessee.

Knowledge

Compared with 2014, the UAW was better equipped with knowledge and more attuned to its membership, allowing it to deploy meaningful information in ways that mattered.

Changing Demographics

Youth

In 2024, the UAW understood the demographic profile of the workers it sought to organize and developed a campaign that catered to their needs and concerns. Since 2019, VW had nearly doubled its production staff and was now seeing an influx of newcomers, many from other parts of the country, and a significant number with prior union experience.[42] Workers believed the new arrivals were key to a reinvigorated campaign: "There's a lot of employees who weren't there who are there now. That's what is spurring the momentum."[43] While railing against the union, even Republican senator Bo Watson could not help but acknowledge that the plant's workforce was now more geographically diverse, stating:

> I hope that the workers from Tennessee will not fall prey to the influence of outsiders who have moved into our area for the incredible economic opportunity we have created here.[44]

These fresh faces were new to the company and trended younger than the general workforce, many of whom were now hedging toward middle age, even if they had been among the original cohort. Steve Cochran, president of the UAW Local 42, acknowledged this demographic shift:

> We have a lot of younger workers in our plant, and the younger generation is by far stepping up big in this union movement.[45]

Another worker expressed a similar view:

> Demographics in the plant has changed since I first started. It's a younger generation that's there and they're more open to the union concept. So they were really willing to listen to us and hear what we had to say.[46]

It was not lost on organizers that younger autoworkers have the added benefit of reduced health-care and pension costs, which weigh heavily on established firms.

The Big Three contracts contained specific provisions tailored to appeal to this young, less experienced workforce. Though there is no guarantee that the UAW will be able to extend the provisions of its Big Three contracts to Chattanooga, certain aspects have had particular appeal to VW workers. At VW, turnover is constant because of the demanding nature of the job and high injury rates, while the competitive nature of the hiring process and management preferences result in recruitment drives that target new entrants to the labor market. This combination of factors means the VW workforce skews young and inexperienced compared to the Big Three. It was, therefore, more than a happy coincidence that the Big Three contracts had delivered a stinging blow to the two-tier system by eliminating the pay differential, boosting beginning pay, and shortening the phase-in for temps—all major improvements disproportionately benefiting the young and inexperienced.[47]

Unfortunately, the UAW's courting of young workers came at a considerable cost. Among the few disappointments in the Big Three contract was its failure to recover the pension plans the UAW had given up during the Great Recession–induced 2008 bargaining round. Pensions had always been a stretch goal given the sheer expense of retiree benefits, but they were a high priority for older workers. Bargaining demands are never a simple zero-sum trade-off, but some legacy employees nearing retirement believed the UAW was courting the next generation and abandoning their cause. Whether or not this assessment is fair or accurate, the UAW's elder neglect made little difference in Chattanooga, where few workers are over forty. In contrast, although undoing two-tier merely healed a self-inflicted wound, it signaled that the union was attuned to the interests of less skilled and more inexperienced workers, whose inconsistent support had contributed to past defeats. It also provided a powerful counterargument to the opposition's depiction of the UAW as staid and out of touch.

Former Union Members
Another demographic shift also benefited the organizing drive. In the past, VW had carefully avoided hiring those with prior union experience. But by 2024, in its frantic desperation to meet production quotas and fill critical positions, VW could not afford to be so picky. Some of these former unionists would go on to play essential roles on the organizing committee. Previous studies of foreign automakers have often pointed to their intensive screening process, which ferrets out those who have previously worked in a unionized workplace.[48] VW might have preferred to hire fewer seasoned unionists, but immediate staffing needs left them with little choice. At VW, there had always

been a minority of workers who had spent time at another southern auto plant. But to the extent that some workers did have prior automotive experience, they were often relegated to peripheral or marginal roles where their impact was minimal. Management seemed to prefer unskilled, impressionable laborers who could be inculcated with "the VW way" from the start. This was a deliberate choice by the company to impose corporate culture and minimize competitor influence. Even when management did hire experienced ex-unionists, it often shunted them to the side, installing them in peripheral departments like transportation or parts rather than assigning them to assembly.[49] But with production ramping up in 2022, management had little choice but to admit experienced workers and former union members to core assembly positions.

Race

The "race-blind" 2014 campaign concealed an implicit racialized and gendered hierarchy that conferred privilege through skill, whiteness, masculinity, and age. By neglecting to build a comprehensive organizing committee that extended into each unit, subunit, and department as well as across shifts and job classifications, the UAW had practiced de facto exclusion, ensuring its campaign would cater to and appeal to the demographically unrepresentative substrata that constituted its primary proponents.

But the plant was not especially diverse during its early years. Some estimated the workforce was 90 percent white in a city that was 29 percent Black.[50] Anecdotal evidence suggests that anti-union sentiment ran rampant in pockets of homogeneity, and research shows that support for unions is significantly higher among Black and Hispanic people than among white respondents.[51] By 2024, the plant had become considerably more diverse, so the organizing committee may have reflected the new composition of the workforce even without proactive efforts on the part of organizers.[52] Recruiting from beyond the Chattanooga region may also have helped, as southerners tend to have less favorable views of unions than people living elsewhere in the country, although post-2022 data is lacking.[53] One worker saw racial and regional diversity combining to dilute the instinctual hostility to unions she associated with white southerners:

> I think because of our great diversity, it's diluted some of that Southern political mentality. And so it's making the conversation easier.[54]

Another concurred, saying the diversity of the organizing committee directly affected the vote's outcome:

> Among African American women, there has been a boost as far as getting it organized this time. I see a big change.[55]

Skill

Previous campaigns had placed skilled machinists in the most prominent roles, forming what the previous organizing director had described as a skilled trades "vanguard." Whatever the merits of this strategy, skilled workers were demographically distinct from the rank and file and thus were poorly equipped to lead an organizing campaign. Though machinists were likelier to have prior union experience, they were older, whiter, more likely to be male, and better paid, and they traveled longer distances to work. Moreover, given the nature of their jobs, they had more physical freedom on the factory floor, which increased their ability to wander the factory and initiate conversations but created considerable resentment. (Most production line workers remain stationary at their workstations for the duration of their shift.) Though skilled trades workers had always been among the most consistent union supporters, they could not parlay these strong pro-union attitudes into broader support throughout the plant.

In contrast, the 2024 campaign deployed a much more comprehensive strategy, elevating unskilled workers to leadership positions and establishing a tightly coordinated multitiered structure of captains to ensure the campaign reached every department and every shift. The UAW doubled down on its bottom-up strategy by building an organizing committee that was representative in terms of skill, gender, and race. Units that had previously harbored pockets of anti-union resistance—such as the paint shop, which, for hygienic reasons, is physically isolated from the rest of the plant—now became bastions of union support.

To be sure, skilled workers continued to feature prominently. Since 2015, Local 42, in its informal guise as a microunit, had persisted at the plant despite legal challenges that prevented formal recognition or bargaining orders from the NLRB. Several Local 42 members, including its president, Steve Cochran, spoke regularly to the press and played an important role in the UAW's public-facing strategy. Cochran's visibility reminded workers that the UAW already had an organized (if unsanctioned by the authorities) presence at the plant. However, behind the scenes, the plant's significant unskilled majority was running the campaign.

Temps

Those with memories of 2014 or 2019 recalled that mass layoffs of temps had deflated the union movement at critical junctures. In both 2014 and 2019, the status of temps was an important factor in the vote outcome despite their exclusion from the bargaining unit. But in 2024, temps could no longer serve as a buffer because the temp agency had been eliminated, making all workers eligible to vote. Moreover, by 2024, the plant had not seen layoffs in three years, and its ramp-up was expected to continue.

Ten Years in the Desert

Perhaps ironically, not all varieties of knowledge aided the UAW's cause. The UAW's loss in 2014 resulted in recriminations, finger-pointing, and general frustration with staffers, who some felt had sold them out or otherwise botched the campaign. Though some continued to support the UAW's ongoing organizing effort, the memory of past failures left a lasting impression that hung heavily over each successive attempt.

However, natural attrition coupled with high injury rates meant that by 2024, relatively few veterans from 2014 remained employed. Save a few stalwarts (including President Steve Cochran), the union's public-facing base consisted of more recent hires. One worker estimated that of the two thousand original hires, fewer than four hundred remained in 2024.[56] Such high turnover may have hurt the drive by compromising its institutional memory, but it also allowed for a new crop that had not suffered the disenchantment and disappointment of 2014. Though VW touted a turnover rate "well below the average for Southeastern plants,"[57] the region's nonunion status set the bar artificially low. (Unionized plants have a much steadier workforce.) Newcomers viewed the campaign with fresh eyes that had not seen the disappointment of a multiyear struggle slipping away.

PTO

If the boldest pronouncements are to be believed, UAW's fate in Chattanooga was a litmus test for the future of the labor movement, the auto industry, the South, and maybe the American economy itself. But for all of the stake raising, the central organizing issue in 2024 ultimately came down to a Christmas holiday.

The demand that featured most prominently in 2024 was an enormously unpopular policy requiring workers to use their PTO during the winter maintenance cycle.[58] The UAW prioritized this issue after conducting solid research about workers' priorities. Workers had already made it clear that PTO was their top concern. Echoing the walk on the boss in 2019 that injected new life into Local 42 after years of quietude, the buildup to 2024 arguably began in 2022, when seven hundred workers signed a formal complaint about PTO changes.[59] PTO remained a major sticking point up through the election (alongside demands for advance notice for weekend work and better provisions for overtime pay).[60]

Though pay might seem a more obvious central demand, especially after the 11 percent "UAW bump," rallying around pay was a nonstarter. A production team member in assembly acknowledged that pay was not the primary concern in 2024, saying, "Money comes secondary in all our conversations."[61] Workers may have preferred earnings on par with those in Detroit, but the fact that pay was already relatively high and improving made it a less salient and weaker agitational issue. A social-media-friendly sizzle reel released by the UAW a month before the vote highlighted PTO, health care, and an end to arbitrary managerial decisions but did not mention pay.

As with the historical shorter-hour movement, calls for sick leave and additional PTO were demands not simply for less work but for *fewer hours with no reduction in pay.* The question of time is often overcoded by the zero-sum logic of *work-life balance,* wherein time spent "living" may have intrinsic value to the worker but is necessarily uncompensated as it produces nothing of value to capital. Breaks and vacations can be prolonged, provided the worker is willing to sacrifice their pay. PTO and paid sick leave turn this logic on its head, providing payment for nonwork. In this sense, both echo what was among the most controversial aspects of the contract campaign: the thirty-two-hour week. Though dismissed out of hand and widely ridiculed, it reportedly emerged from membership surveys and resulted from pandemic-related revaluations. One worker noted, "Having a good wage is important, but workers also need to be able to enjoy the fruits of our labor."[62]

Time, even more than money, was once the main question at the core of the labor movement's remit. It has often been noted that for all of COVID-19's death and devastation, it recast the value of time and forced a revaluation of leisure. The pandemic highlighted the importance of flexible hours and the need for safeguards against unexpected illness.[63] During the stand-up strike, the UAW's demand for a thirty-two-hour workweek with no loss in pay had always been one of its more controversial goals. It was always a long shot, and it seems to have made little traction at the bargaining table. But the UAW was serious in its commitment to shorter hours, even sending Fain to testify before Congress to support Senator Bernie Sanders's bill to make the thirty-two-hour work week standard.[64] In the same speech, Fain suggested that the fight for a shorter workweek could be a part of the UAW's next contract campaign.[65] Though the thirty-two-hour workweek was a nonstarter, workers became eligible for eighty hours of paid parental care leave under the Big Three contract.

Yet there may be other reasons for the UAW's silence on pay. Capital flight is driven by two interrelated pressures: capital's incessant demand for low-cost labor and its desire to maximize efficiency by reorganizing the terms and conditions of production in locations where worker resistance is weak. But globalization struggles often center on speedups, quality control, and output while sidelining fundamental compensation questions.[66] Although each new construction factory represents an increase in machine sophistication and a new potential for productivity, the driving force behind global expansion is the low wages that can be paid in VW's far-flung facilities. In this sense, struggles over relative surplus value distract from more basic questions of absolute surplus value.

Sick Days

Sick days were among the workers' biggest concerns, second perhaps only to PTO. Workers noted that injuries were rampant on the line: "I have never worked for a place where I have seen so many people damaged and broken."[67] As with PTO, workers objected not only to the quality of health care but also to its seemingly arbitrary character, which left them unable to budget:

> Right now, VW has total control over our health coverage, and frequent changes leave workers with unexpectedly high costs. This has caused some workers to forego important medications.[68]

Frequent injuries resulting in extended leaves of absence also impacted the organizing committee, making it difficult to maintain continuity:

> Among the [organizing committee], stories of maimings were frequent. We have lost activity from . . . organizing committee members because they got bodied at work.[69]

Though it received less attention, sick leave was another sticking point (workers had none). As one worker put it:

> In mid-February, quite a few people came in sick because they didn't want to get penalized. And then of course they got everybody else sick, and then we have a whole bunch of people out. Everybody's getting disciplinary action and losing bonuses because they're sick and can't come to work.[70]

Workers wanted the ability to credit their PTO beyond the mandatory holiday and to eliminate the impossible choice between canceling a planned family vacation or working while sick. The lack of sick leave, rigid PTO policy, and stiff penalties for absenteeism created a culture of fear. Even though the issue could have been resolved at relatively low cost to the company, a fateful decision was made to kick it down the road for five years. While the drive was overdetermined, this decision, more than anything else, set in motion the chain of events that led to the eventual victory in 2024.

Learning

Learning flowed naturally from the innovation, motivation, and knowledge that the new UAW brought to the table. On measures of militance, inclusiveness, transparency, and democracy, the UAW fared better than in 2014 or even 2019.

Militancy

Fain's militant posturing is a dramatic about-face for a union that once built an entire automotive division (Saturn) as a monument to labor-management cooperation. Militancy necessarily has a contradictory character. Cutting past the tough talk, unions are burdened with the responsibility to sell their members' labor for the best price. Much as Fain styles himself in the image of an old-school labor militant, he is ultimately responsible for managing a national organization with $1 billion in assets that moonlights as a social service agency, recreation center, and HMO. In the 1930s, militancy often involved testing the limits of the law, a daunting proposition for a union whose lawbreaking had recently earned it a full-time federal monitor. Fain, the man who famously chastised billionaires on national TV, was also responsible for managing an organization with a real stake in preserving the status quo, not to mention a pension fund that bankrolled those same billionaires. For a not-insignificant group of prospective members, the primary draw of the UAW was the long-term security it promised, not lunch-pail socialism.

As the UAW had already learned, striking is a risky organizing strategy. Though the GM strike in 2019 involved nearly as many workers as the selective strike in 2023, it seemed more an act of desperation than a demonstration of strength. This time, in a further demonstration of the union's power, Stellantis and GM agreed to pay each striker $110 a day for their time on the picket line, on top of the union's strike pay.[71]

The tone and tenor of the UAW's 2024 campaign was more impassioned and combative. Previously, the union had avoided antagonizing the company and had taken a soft-pedal, low-key approach to organizing, but the UAW came out of the gate in 2024 with a renewed fighting spirit. It was highly motivated to maximize gains for members rather than settle for weak or even concessionary contracts, as had been the case over the last two contract cycles. As it had done in 2019, rather than relying on the company's good graces, the UAW pulled out all the stops, calling out VW's abuses. However, separating 2024 from 2019, the 2023 Big Three negotiations revealed it was willing to take job actions, up to and including strikes, to secure these gains. Under an agonistic labor relations model, a union that enters contract negotiations prepared to concede has already lost the battle. This time around, rather than coddling the company, the UAW used a back-to-basics organizing approach that echoed its storied history of militancy. VW workers responded to that message. Having witnessed the limits of labor-management partnership, workers were no longer willing to compromise:

> People [are] like, yeah, we don't want to hold hands. . . . What you're doing is wrong. And we're the workers, so yeah, we're not your friends.[72]

Community Engagement

In 2014 and 2019, the opposition had presented itself as a homegrown, organic alternative to jet-setters and carpetbaggers opposed to the southern way of life (even though it was instigated and funded by deep-pocketed outside agitators). Bereft of community support, the UAW could rebut the opposition's arguments but had no counterattack plan. Community support is symbolically important, as it resists efforts to portray the union as a third party and gives organizers an added measure of legitimacy. But as Ganz observes, community supporters also bring a tangible benefit to organizing drives through their command over local knowledge.[73] Against efforts to portray Chattanooga as a timeless monument to unencumbered capitalism, a local community organization launched a public education project, reminding anyone willing to listen that 40 percent of Chattanoogans had been union members as recently as the 1960s.[74] (This information was available to the UAW and well known to any serious historian of the South, but it meant more coming from a group with local roots whose core constituency included veterans of these struggles.)

Out of the Shadows

Received wisdom holds that union campaigns should initially operate under the radar to avoid confrontation and any publicity at all, at least in the early phase of the campaign. Unions should avoid going public until the last possible moment to catch the opposition by surprise. Conventional union drives are covert, building support gradually and postponing a formal announcement until the last possible moment, hoping to fly under the boss's radar and denying the opposition time to mobilize.

In this regard, the UAW broke sharply with conventional wisdom. Invigorated by the stand-up strike, workers felt emboldened to mount a public strategy from day one. As Alex Press has observed, the UAW "flip[ped] the traditional order of operations" by openly announcing its plan to organize the plant even before it had boots on the ground in Chattanooga.[75] This strategic shift resulted in part from a changing context, but it signaled a shift to a more visible style of organizing that did not shrink from confrontation. No less a tactician than Jane McAlevey endorsed above-ground organizing campaigns in her late work.[76] Under normal circumstances, public campaigns risk emboldening the opposition, but with momentum on its side, the UAW could afford to take chances. As it turned out, the gamble paid off. Going public early allowed the union to harness the power of social media to broadcast its message from a virtual soapbox and brazenly distribute union authorization cards on the floor without wasting precious time. By the time the opposition reemerged, the UAW already commanded an indestructible supermajority, and the vote was little more than a perfunctory affirmation of the overwhelming consensus.

In the UAW's calculations, the advantage gained by carrying forward momentum from the stand-up strike was more than balanced by the risk of tipping off the opposition. With a public strategy, workers were able to accelerate their campaign, more than making up any time they had lost with an early announcement. Workers reported that organizing happened rapidly once the organizing committee had begun making headway. Within three months of going public, the UAW would file for an election. By going public immediately after its November victory, the union put forward a narrative explicitly connecting its watershed deals with Big Three companies to prospective contracts at transplants, raising expectations and generating enthusiasm in Chattanooga. Indeed, after its public announcement, the union proceeded to surpass organizing milestones at a record pace, collecting cards from 50 percent of workers within two months and declaring supermajority support and filing for an election just four months after signing the Big Three contracts. At that point, buffeted by new rules from the Biden administration allowing for an expedited election process, the vote could be called within three weeks. The six-month interval from the union's public announcement in November to the vote in May gave the opposition plenty of time to ramp up, but it was counterbalanced by the propulsive force of the stand-up strike—an effect that might have dwindled had the union delayed its announcement until the spring. Comparing 2014 to 2024, one worker recounted:

> The people that were pushing for the UAW, it was like we were part of a secret society. We had to be real hush-hush about it because we didn't want to get in any trouble with HR because we said the word "union." So it was real hard for us to get the word around to our co-workers because of the pressure from management to not have any type of discussions whatsoever about the UAW in the plant.[77]

But in 2024, under the new NLRB rules, workers gained the right to speak freely about the union during company time. According to some, this was a decisive factor:

> By us being able to communicate a lot more, it allowed us to educate our coworkers to be able to give them more information about what UAW is.[78]

To comply with these regulations, supervisors were obliged to read a memo declaring unequivocally that workers were free to organize without fear of retaliation. This allayed any lingering fears workers may have had.

The UAW's public strategy also benefited from changes in labor law. Previously, the union had sought to keep its intentions under wraps until the last possible moment, knowing that the NLRB could postpone elections for months, granting the opposition the gift of time. Under new NLRB rules, elections had to be held within two weeks of a majority of employees signing cards, barring voluntary recognition. Therefore, the UAW could afford to go public early, knowing its opponents would have little time to mobilize.

As noted earlier, themes of trust and deceit figured prominently in previous defeats. The union, its supporters, its opponents, the company, its agents, and politicians were all branded liars at various points. By conducting its affairs openly, the UAW signaled that this campaign was different. Indeed, by casting aside years of received wisdom and announcing its intention to organize VW before it had even achieved the 30 percent benchmark, the UAW showed members that supporting the union was not a shameful secret but a source of pride while also implicitly discrediting opponents who sought to portray the union as a clandestine conspiracy.

After spending much of 2013 patiently awaiting the possibility of card check while the opposition staged a multifront attack and steadily picked off its supporters, the UAW had learned that time was its worst enemy. In 2019, it sought to move swiftly from the filing to an election, but legal challenges defeated these efforts, and opposition once again used the interim delay to flip the result. In 2024, new NLRB rules shortened the filing window, allowing for an election only weeks after filing.

Member-Centered Organizing

Perhaps the most notable characteristic of the 2024 campaign was its bottom-up orientation. By all accounts, a greater share of the organizing was performed by members operating within a complex multitiered structure of delegated authority and responsibility. The UAW developed a 300-worker-strong volunteer organizing committee from a pool of roughly 4,300 eligible workers. Put differently, approximately one out of every fourteen workers was a volunteer organizer. These numbers do not include other union supporters who signed on to support the campaign in a more limited capacity. Given the stakes, even in the context of a drive notable for its public character, not all workers necessarily felt comfortable endorsing the union publicly or taking highly visible roles. Still, there were roles for workers who preferred to assist from behind the scenes.

As noted, the 2014 drive centered on skilled machinists, who were prized for their mobility but were atypical of the average employee in terms of pay, qualifications, and demographic characteristics. The 2024 campaign drew more heavily on the large majority of the workforce in fixed production ti-tles, often tied to a stationary workstation. While production workers may not have had the synoptic perspective of a machinist, they could forge deep-er connections with small groups of proximate coworkers. Combining mo-bile roamers with production workers allowed the union multiple points of contact with fence-sitters, which it could use to help persuade these ambiva-lent employees through targeted conversations.

Member-centered organizing is not new to the union movement, but it was to the UAW, which had favored staff-centered campaigns even as it uneas-ily adopted the organizing model as its membership dwindled. The indi-vidual responsible for dragging the UAW toward a member-focused approach was Brian Shepard, a veteran of the union movement with experience at SEIU, one of the unions that has pursued organizing most aggressively. Though Shepard came to VW with a good sense of best practices known to win cam-paigns, he deferred to the workers on many matters of strategy. For example, when he broached the contentious issue of house visits, which some had fault-ed for the loss in 2014, workers insisted that Chattanoogans would not take kindly to visitors, especially if those who appeared on the doorsteps were staffers. (For reasons of time and practicality, worker-organizers are under-standably reluctant to commit to making house visits in the hundred-mile swatch of Appalachia that constitutes the plant's regional footprint.) Rather than implement house visits over worker objections, Shepard relented. Whether or not the concern was merited, ceding on house visits gave work-ers a sense of control over the campaign.

Rather than impose an agenda from above, the UAW allowed workers to take the lead in identifying their issues. In the past, the UAW had often high-lighted the pay differential between Detroit and Chattanooga. But that mes-sage had never been especially persuasive in Chattanooga, where VW pays well above the median income. By 2024, progressive pay increases had lifted VW even further above the regional standard, so pay no longer resonated (if it ever had). Instead, health care and PTO—both concessions that, if granted, would likely cost the company less but that had real traction on the shop floor—became the flagship issues of the campaign.

Previously, the UAW had demonstrated a lack of commitment to formal democracy, running top-down, staff-driven campaigns, especially at VW. In 2014, workers felt they had been sold a bill of goods because the UAW went over their heads and cut a deal with the works council before the vote took place. By 2024, UAWD's rank-and-file rebellion had dismantled the AC and replaced it with a new generation of reformers, switching to a one-member, one-vote model for the first time in the organization's living mem-ory. This renewed commitment to formal democracy meant that workers at any newly organized shop would have a louder voice and more avenues for participation. Corruption in the organization was all but eliminated, both by the jailing of top officials and by the introduction of a different culture within the organization that was more transparent and, therefore, less vulnerable to abuse and more easily overseen. Perhaps reflecting this cultural shift, 2024 made greater use of worker-organizers compared to 2019 and 2014, with hundreds of workers beyond the formal organizing committee volunteering for these roles. Though union staffers provided the organizational infrastructure, staff preferred to relinquish control over the day-to-day work of organizing. Reflecting the centrality of members, the single most prominent issue highlighted in campaign materials was the PTO policy, even though, in dollar terms, it was worth less than the union-rate wages members were likely to demand. From the standpoint of the rank and file, the seeming arbitrariness with which the company made decisions was at least as salient as the gross undervaluation of nonunion autoworkers' labor.

In 2019, the UAW had prioritized public relations, seeking to match one-to-one the opposition's carpet-bomb media strategy, blanketing the town with propaganda. In 2024, the UAW dedicated its limited resources to inside organizing, with only selective advertising on social media targeting workers. Additionally, the UAW made sophisticated use of text messaging, social media, and digital organizing to supplement face-to-face work. Revamped in 2014, its communications department could now produce professional-quality viral content in-house.

Finally, by 2024, the UAW had reaffirmed its focus on organizing, now backed up with dollars. In the past, the UAW had tried to substitute expensive ad buys for on-the-ground organizing. Steeped in a culture of corruption, the UAW was a latecomer to the so-called organizing model pioneered by SEIU and Hotel Employees and Restaurant Employees Union (HERE). This time, it avoided a major media spectacle, focusing on organizing, not theatrics. Effective organizing requires skill, expertise, and money.[79] According to an analysis by the investigative reporter Chris Bohner, the UAW's $40 million commitment to new organizing funds represents a doubling of its organizing budget compared to 2021 (the most recent year for which data is available, following new disclosure rules).[80]

Recounting the differences in organizing style between 2014 and 2024, one worker said:

> It's much more of a supportive role . . . helping guide and train and develop more organic leaders from within the membership, the rank and file, to keep taking the fight to the next level.[81]

The previous leadership had kept members in the dark during negotiations, arguing that the complex dealings would be compromised if members were allowed to see how the sausage was made. One worker said:

> My last two contracts were, "bargaining is going fine".... Then eventually, "here's a s—— contract, we want you to vote yes on it."[82]

Fain believed it was in the union's interest to promote transparency. Weekly updates via Facebook steered clear of minutiae but offered a top-level running commentary, encouraging ongoing internal debate well before the tentative agreement was distributed.

Works Council

The 2024 campaign was not a complete break with the past. There is some evidence that the works council exerted influence behind the scenes to pare back the company's anti-union response. For example, the works council wrote a letter threatening "severe consequences" if the company deprived workers of their right to organize.[83] However, the relative importance of the works council's intervention is a difficult question to address empirically, given that other contributing factors in 2024 and similar promises in 2014 had little impact. Outward signs suggest that while the works council was still involved in the drive, it no longer commanded the central role it once had. Years earlier, the IndustriALL Global Union had suspended its voluntary global agreement with VW in response to the company's refusal to negotiate with the machinists.[84] Since the GFA had been declared null, the works council and its partners no longer had a company-specific weapon in their arsenal.

In 2014, the works council had dominated discussions, even though its influence was often misunderstood. It functioned as a structuring absence, shifting the discussion away from the merits of unionization and toward a mysterious organization with a long German name that seemed to function as the union's proxy. In 2024, workers rarely mentioned the works council, except when pressed, and did not perceive it as having a significant influence. True to form, rather than exert its institutional power at VW headquarters, the works council elected for a series of scattershot exchanges and publicity stunts that were limited in risk and impact.[85] Even if the works council was responsible for the company's restraint, from Chattanooga it appeared to be only one of a chorus of sycophants and hangers-on, not the main act. Without the works council's mediating influence "run[ning] interference" between the company and the union (as one worker put it),[86] workers had the opportunity to evaluate UAW on its own merits.

Conclusion

Even when organizations engage in learning practices, the benefits may fail to materialize in the presence of organizational rigidity and inertia. In 2019, the UAW was calcified and fixed in its habits to the point that it could not pivot or adjust its strategy to reflect changing conditions. Still reliving the trauma of 2014, it found its path forward darkened by the shadow of the past, its yearning for victory haunted by a desire for vindication. Just as the obsessional neurotic organization must pass through *and beyond* analysis to avoid the repetition of previous pathologies,[87] the UAW had to stop endlessly replaying and reconstructing 2014 to move ahead. If those who fail to learn from history are doomed to repeat it, those who fail to escape the clutch of history are condemned to relive it.

Attacks on the social wage and the general degradation of labor over the last half century have understandably produced a desire to recover what has been lost. Innovation exchanges the quotidian grind of trial and error for the uninhibited musings of the social imaginary.[88] But unions must exercise caution to avoid a politics of nostalgia that seeks a simple return to the supposed "golden age" of the postwar social contract. While innovative unions are not oblivious to previous experience, they are disinterested in endlessly rehearsing the past. The revitalization of the labor movement demands an anticipatory future orientation and a willingness to forsake old habits, comfortable though they may be.

8

Solidarity's Intimacies

Topologies of Labor Power

Extending beyond our case study, this chapter stages a critical intervention in the emergent field of new global labor studies (NGLS), questioning and problematizing some of the core strategies that have come to define emergent research on labor's "new internationalisms" in the twenty-first century. As noted in Chapter 4, an inordinate overemphasis on transnationalism lends itself to paper agreements and high-level alliances that fly above the plane of the local, never descending to the often-messy terrain of grounded struggle. Thus, the scalar challenge is twofold: successful transnational campaigns must be both *extensive*, expanding across and transcending territorial limits, and *intensive*, cutting through and reconfiguring the topological planes that link putative social actors. Importantly, both strategies overflow the nation's boundaries, whether by expanding beyond its confines or burrowing within its folds and undulations.

In conjunction with Chapter 7, this chapter locates the root cause of the UAW's turnaround in the union's regime change and organizational transformation. The democratic revolt of 2023 was a systemic shock that reverberated throughout the union, reshaping both its formal composition and its informal organizational culture. In 2014, rather than take calculated risks, the UAW had pursued the path of least resistance, settling into patterns that had already proved ineffective. By 2024, it was open to experimentation, unafraid to leave its comfort zone and assume the risk of the unfamiliar. Apart from reshaping union strategy, the UAW's reinvention also had meta-organizational implications, repositioning the union within the social field and reorienting its alliances and connections. Just as rank and filers dismantled autocratic rule in Detroit, they had little tolerance for the rigid hierarchy of established social governance regimes. No longer beholden to a corporatist mindset, the UAW shifted its focus to the automotive sector more broadly, highlighting the similarities between firms rather than stressing their differences. As the code of deference and submission faded, power diffused throughout the organization, and workers were free to develop organic connections, which gave rise to intimate and affinitive ties. The command-and-control administrative state gave way to inclusive, traversal, and circulatory flows of information and resources. In short, the systemic shock that had reshaped the union from within unleashed a recombinant unionism: an assemblage of shared interests constructed through bricolage—or an *amalgamation* in the original sense of the word.

Seeing like a (Labor) State(sman)

Extending beyond our case study, this chapter stages a critical intervention in the emergent field of NGLS, questioning and problematizing some of the core strategies that have come to define emergent research on labor's "new internationalisms" in the twenty-first century.[1] The UAW's previous efforts to build transnational alliances mirrored its top-down, socially engineered, and firm-centered organizing style. Its fitful gestures toward transnationalism suffered from a misguided approach to solidarity. Holding connection and difference in tension is a complex challenge. But retreating toward the comfort of a defensive neonationalism is no solution—nor is delusional overconfidence in the hypernationalist myth of the Other.

In the final analysis, the UAW's global aspirations were also hampered by the murky legal status of any cross-border labor campaign. Labor law is established and enforced by and within the jurisdiction of states, so any move toward labor transnationalism requires reconciling divergent policies across national borders and between systems that may have fundamentally different approaches to industrial relations. Though the globalization of capital often requires interfirm coordination,[2] workers rarely develop congruent networks of struggle and instead remain tied to local labor markets.[3] Even more damning, GFAs derive their moral authority from the ILO Conventions on Freedom of Association and Collective Bargaining, which have yet to be ratified by the United States. Signatories to GFAs often argue that when local legal requirements set a lower bar than a GFA, only the less restrictive local requirements apply. Since U.S. labor law does not meet the minimum international standards set forth by ILO codes and effectively tolerates violations of both freedom of association and the right to bargain collectively, many companies argue that GFAs are essentially meaningless for those doing business in the United States. Indeed, much of what passes for transnationalism is more appropriately termed *supranationalism* in that it maintains and even cements national boundaries while gesturing furtively toward a supposedly transnational ideal. In the process, it does little to challenge the schism between North and South that incentivizes globalization in the first place.

At minimum, a globally oriented labor movement must oppose the idea that the nation-state is the sole arbiter of its fate by radically reimagining the categories through which solidarity might emerge. According to Rebecca Johns, "Dominant conceptions of internationalism assume a set of closed boxes, tidily bounded by nation-states out of which workers reach hands across seas or national borders. The boxes themselves, though, straddle nations."[4] How can one reconcile the global with the local without giving either short shrift, all while escaping the statist trappings that preclude transnational solidarity? Global solidarity cannot be achieved through the false unity of tethered division, as if transnationalism is nothing more than a compilation of nationally bounded struggles.[5] Striking a successful balance between the global and the local requires not simply a rescaling of labor's organizational infrastructure but a careful assessment of the social relations that compose that structure. For all its historical significance and inspirational power, the notion of a transnational labor movement is less an empirically observable phenomenon than a speculative object of intrigue. Perhaps for this reason, the burgeoning field known as NGLS remains undertheorized, underconceptualized, and lacking many useful actually existing models.

Transnationalism is based on the idea that unions might exceed, escape, or transcend the strictures of context. The *trans-* in *transnational* suggests both *transversal* (as in crossing) and *transcendence* (as in overcoming). This ambiguity lies at the heart of labor's transnational project, which aspires to surpass the nation even as it remains dependent on the nations that constitute it. Thus, as much as NGLS might imagine itself flying above late modernity's jigsaw of nation-states, it remains surprisingly flat-footed. The logic of transnationalism is often additive, content to combine bounded national-level struggles rather than exceeding their limits.

Even as the transnational turn fundamentally reshaped the humanities, NGLS has accommodated these important developments belatedly and reluctantly, if at all. With notable exceptions that only prove the rule, NGLS takes as its starting point the existence of clearly defined nation-states whose labor regimes are coextensive with national borders. While the humanities have shifted their focus to diasporas, migration, crossings, borderlands, hybridity, creolization, hyperspaces, and displacement, NGLS maintains the nation as its unit of analysis and basis for comparison. It is perhaps telling that NGLS insists on the formulation *transnationalism*, which implies that a contingent, tendentious process is ultimately reducible to a settled state of affairs, the better to be prodded and scrutinized by experts. But the transnational project is sapped of its dynamism when presented in its hypostatized form. As the experience of the Third International reveals, the transnational project differs from nationalism in both type and kind, even if workers renounce their country and hoist the flag of the global proletariat in its place.

A close examination reveals that NGLS is still wedded to nationalism. The transnational question, as posed by NGLS orthodoxy, is whether a given country's "native" culture and policy mores can be exported offshore. The contingent practices and negotiated spatialities implied by transnationalism are lost in NGLS's emphasis on the fixity of place, the imprimatur of context, and a binary understanding of difference. Transnationalism is something more than the sum of the nationalisms it traverses. Methodologically, the preference for the comparative case study authorizes cartoonishly unsophisticated portrayals of context, the better to juxtapose two fundamentally dissimilar examples. Even when dealing with *transnational* corporations (as opposed to relatively place-bound *multinationals*), NGLS imagines that these firms inherit character traits from their "parent" (headquartered) countries. National settings become defining features of the landscape as companies and unions are marked and imprinted by the determinate social environments they inhabit.

Scale

NGLS has lagged behind adjacent disciplines by failing to reckon with space. Though NGLS tends to reify the categories *local* and *global*, the global is often little more than an accumulation of locals arrayed strategically to appear as something more than the sum of their parts. Meanwhile, the local traces its existence to nationally bounded regimes on which it depends. Even "mainstream" geography increasingly understands the global and the local not as fixed categories but as heuristic devices.[6] But NGLS scholars continue to treat "the local" and "the global" as relatively unproblematic descriptors of known scalar dimensions. NGLS must be brought into dialogue with economic geography by calling into question key assumptions about the composition and structuration of space, problematizing key assumptions about the nature and site of power. In short, NGLS remains wedded to the outmoded logic of place and space. The following section unpacks and deconstructs the spatial encoding of labor economics before proposing an alternative conceptual model.

Hyperglobalism

Classically, labor's main problem was how to exceed the strictures of its nationally bounded interests and achieve its transnational mandate by rising to the scalar level of global capital. Capital was content to present foreign firms as out of reach and irrelevant to the day-to-day struggles of American auto-workers. The opposite also held: American workers doubled down on their entrenched nativism, presenting Japanese and European competitors as, at best, an outside threat rather than an endemic symptom of a problem that implicated all firms irrespective of origin. Therefore, labor's task was to *scale up* and transcend the confines of the local to embrace the global instead.

In this standard telling, "the global" is coded as the realm of capital, which traverses continents in search of markets without allegiance to any particular site. Conversely, "the local" is the site of labor, grounded on the shop floor and embodied by discrete individuals with familial ties and es-tablished roots. But globalization has undermined the stability of the "em-bodiable topos,"[7] emptying these spaces of their terroir and undercutting any vestigial sense of place that might have once formed the basis for action. Both labor and capital have been reduced to unhelpful caricatures that exag-gerate capital's placelessness—when it remains highly dependent on states and local regimes—while understating labor's mobility in an age when the global proletariat is defined by its migratory status. Moreover, this rendering suffers from the confusion of space and movement, where mobility is imag-ined to diminish the importance of place.

According to this trope, capital is impressive but harmful; labor is meek but virtuous. The global reasons; the local emotes. At times, labor is explic-itly coded as feminine (independent of whether workers are actually female). Labor's weakness results from its provincialism: capital looks down on labor from magisterial heights, but labor cannot respond in kind. The global is the site of programmatic logic; the local is the place of impulsive contingency. Politics can only occur at the level of the global, whereas the local is the site of the social. Facing such profound asymmetry, labor must "scale up" and ascend to the grand scale of capital to defend its local interests against cap-ital's overreach.[8] Confined as it is to nationally bounded regulatory regimes, labor views capital's enormous expanse and physical mobility with a certain jealousy tinged with ressentiment. Since the birth of the union movement, labor has aspired to match capital's grand scale, a project that seems ever more pressing today as the largest companies now encompass truly global dimen-sions.

Hyperlocalism

Critics of the hyperglobalist position have observed that building transna-tional alliances (*extensification*) seems to contravene the *intensification* of the "locally" based initiatives from which unions derive their power. For these scholars—who echo the alter-globalization discourse of the early 2000s—

the political project becomes one of rescuing place-bound labor from the placelessness of a capital's overarching power. In this view, globally oriented movements risk effacing rank-and-file initiatives and represent a threat to internal democracy as such.[9]

Andrew Herod is surely the best-known figure behind the ascendant localist position (and later, the midrange compromise). Herod's influential work offers an important corrective to narratives that foreground global alliances while downplaying the importance of local struggle.[10] But as he acknowledges elsewhere, the fetishization of the local comes with its attendant risks and exclusions. Examined in its totality, Herod's work splits the difference between a localist and globalist position, arguing that both are necessary depending on the circumstances.

Adopting an even stronger localist stance, Katy Fox-Hodess claims that local-to-local shop-floor coordination is the best predictor of successful transnational coordination.[11] For Fox-Hodess, global labor solidarity must avoid the inexorable tendency toward centralization, which is anathema to the small-scale alliances she valorizes. In doubling down on her localist position, Fox-Hodess positions herself against Herod, whom she sees as treating bottom-up and top-down campaigns as equally efficacious depending on the circumstances. But her proposal is also constrained by the particularities of her case study—which draws on strategically placed workers in the longshore sector, who have long wielded disproportionate power relative to their numbers. The classic exemplar of effective, substantive global campaigns has long been the mythical dockworker. But it is not clear that this example is meaningful outside this highly specific industry with its unique characteristics. The endless rehashing of dramatic "success" stories like Fox-Hodess's dockworkers hardly presents a workable model. For example, in the case of autoworkers, though they possess considerable structural power, they have not unleashed it in recent memory, and it exists mainly as speculative potential.

In apparent opposition to the romance of the global, some scholars have mounted an equally problematic defense of the local. Just as an overemphasis on the global may distract from local affairs, the romanticization of the local can be equally debilitating. At times, scholars overeagerly defend the local in the name of preserving indigenous customs and values against the bland leveling force of the globalization machine. In light of this criticism, Herod has modified his position and now cautions against hyperlocalism:

> Defense of the local to preserve traditions and jobs is often one side of a coin whose obverse is rabid xenophobia; such localism may result in a parochial labor politics designed deliberately to protect particular spaces within the global economy at the expense of workers located elsewhere.[12]

This defensive posture may be seductive, but it does little to stem the inevitable erosion of place-bound folkways and much to harden existing obstacles to transnational unity. Understandably, workers might cling to what little is identifiably their own in a campaign as the world drifts inexorably toward cultural homogenization, but the defense of the local on strategic grounds is more persuasive. Given that the global economy is a deeply striated field in which power is unevenly distributed, workers in strategically located sites might exert disproportionate leverage, extracting concessions that exceed their numerical weakness.

Still, the *up* in Herod's "scaling-*up*" is an unfortunate turn of phrase, creating a schematic hierarchy in which places are situated beneath the practice and processes of solidarity, which can only occur in the untethered realm of the universal. Once again, the transnational is cast as the space of abstracted universals; the local is parochial and sensory. Concurrently, scalar metaphors suggest that places are unaffected by solidaristic processes, overlooking the fact that "places" are made by and through their connections as much as their positions.

Among other problems, this understanding is profoundly antidialectical, viewing labor as subsumed within rather than constitutive of capital.[13] Even if theorists contend that the local is constitutive of the global or that the local imbues the global with the energy and passion it so sorely lacks, the global retains its status as the master category while the local remains subservient and subordinate. As with all antinomies that counterpose a privileged and denigrated category, the relation between the local and the global is characterized by a jealous *ressentiment*, in which labor, despite its contempt for capital, seeks to emulate capital's sweeping influence and multinodal coordinated strength. Thus, labor is caught between the romanticization of the global and the fetishization of the local. Additionally, the view of capital as free floating and indifferent to place and space is not altogether accurate. Today, capital staves off critics by striving to embed itself more securely in the local environment, while labor continues to bargain at the local (national?) level. Article titles like "Finding the Local in the Global" speak to this quasi-dialectical tension, in which authors tacitly recognize the impossibility of neatly differentiating between the false taxonomies of scalar geography.[14] Thus, the macro is juxtaposed with the micro, and actors must strive to achieve the appropriate balance between these two countervailing tendencies.

Moreover, the growing call for "local" solutions has received pushback from proponents of centralized coordination. A series of notable voices have expressed profound pessimism regarding the future of bottom-up transnationalism. For Michael Burawoy, the evidence for local-to-local cooperation is sparse, and the examples offered by its defenders point to exemplary one-offs without lasting consequences or replicable outcomes. These exceptions prove the rule to the point that internationalism as a political project remains aspirational.[15] Or, as Gay Seidman writes, "There is no sign that their 'small transformations,' or better their small perturbations, are more than an adjustment to capitalism."[16] Yet local initiatives, however virtuous, must "scale up" to effectively confront capital's enormous reach. Successful examples of cross-border labor alliances generate monographs, even as most struggles remained contained within a nationally bounded regulatory framework. The nation retains its status as an effective container for labor's strategic aspirations and tactical logic, even amid evidence that other scalar units have begun to displace it.

At times, it is imagined that divisions can be overcome through cross-border charrettes and prearranged meetings between workers employed by the same company, as if the main barriers to solidarity were mistrust or prejudicial attitudes.[17] Even transnational alliances have an important face-to-face dimension. However, in a sign that the global division of labor cannot be reduced to a series of interpersonal relationships, such cross-border charrettes have had mixed results.[18] A materialist reading of history suggests that even the ugliest manifestations of xenophobic protectionism have their basis in economic reality.

Middle Ground

With the global labor movement at a crossroads, it confronts two countervailing impulses: scaling up and scaling down. In her reflections on labor's fight over NAFTA, Tamara Kay argues that union leadership bears principal responsibility for the failures of cross-border cooperation, but also observes that such a project is impossible without buy-in from the rank and file.[19] In this sense, labor's continued relevance in a global economy depends on its ability to overflow its domestic confines and its willingness to reinvigorate its membership and foment grassroots insurgency. Herein lies the classic paradox of scale: At first blush, scaling up might seem all but imperative for a labor movement that seeks to escape the confines of bounded, fragmented, piecemeal struggle. But today, the movement for corporate accountability confronts the opposite challenge: maintaining its global ambitions while scaling down to ensure its relevance and applicability at the point of production. The desire to surpass is often accompanied by an appositive, introspective drive, a pulling back. For most authors, this is accompanied by the caveat that such initiatives can only be fragmented and partial. If macroinitiatives are characterized by their synoptic, totalizing reach, microinitiatives are quaint, contained, and shrunken to human scale. The TNC, by its very name, implies an expansive organization, yet TNCs operate simultaneously at multiple scales.

Newer scholarship has moved toward a conciliatory détente between extreme localist and hyperglobalist positions. For David Featherstone, the solution to this impasse lies in a "stronger" conception of internationalism that implies no hierarchy or priority between the local and global realms:

> Scalar levels do not have a linear or diachronic relationship; they do not come one after the other, but recombine (or can recombine) in multiple ways and directions.[20]

While still founded on a conception of local and global as separate, distinct, and fundamentally divergent, this model allows for a more expansive notion of scale in which particular levels are not granted a priori epistemological privilege.

Synthesis

So far, this chapter has identified twinned scalar and relational quandaries that together forestall the possibility of a viable transnational labor initiative. At one extreme is the Harveyite globalist school, where capital establishes the conditions for resistance in advance.[21] Successful resistance can only occur when labor "scales up" to the level of capital's already-constituted transnational dimension. If this view remained preeminent through the 1990s, it has since been largely supplanted by an alternative position, in which scholars offer the appealing, if somewhat romantic, view that hyperlocal small-scale activism might one day blossom into a web of resistance. According to this logic, labor succeeds not by mirroring capital's global reach but by "scaling down" through local-to-local shop-floor coordination. Finally, some scholars have tried to broker a midrange settlement, reconciling the global and the local while preserving their underlying tension.[22]

All three positions have been exposed for their weaknesses. Resistance that starts and ends in the realm of the global entrenches power, generates paternalistic and condescending social relations, and risks the demobilization of its local component. Conversely, local struggles alone, guided solely by altruism or charity as a substitute for solidarity, will become marginalized or irrelevant without the connective tissue of alliances and coalitions. However, splitting the difference through grammatical contortions such as "glocalism"[23] or "grounded globalism"[24] does not resolve the underlying tension and risks logical incoherence.

Topology

This section reviews research on topology, which offers an alternative to topographical conceptualizations of space and place, confounding the categories that undergird the topographical project. Though labor geography is deeply concerned with the exercise of power across space, both power and space suffer from a lack of theorization. Power is often presented in its Newtonian form—a force contained within known entities and exerted on others. Space is likewise presented in topographical format, a mappable terrain with peaks and valleys distributed across a planar region with immutable distances and transects that can be measured in advance. To be blunt, these conceptions of power and space are somewhat quaint by present-day standards of political theory and theoretically informed geography. This is not to suggest that NGLS must embrace recent theoretical developments simply to remain *au courant*, much less to legitimize itself in the eyes of high-theory gatekeepers. NGLS's commitment to solving "real" problems on the ground is admirable, and the field cannot and should not shed its applied orientation. However, to the extent that trite and antiquated conceptions of space and power no longer describe reality—if they ever did—the lack of theoretical engagement may be holding the field back.

Though his class politics are milquetoast and his notion of political action too often hinges on an Arendtian idea of shared responsibility, John Allen's work on topology will prove helpful here. Allen's pioneering work on power topologies contends directly with space rather than accepting territorial boundaries as given.[25] For Allen, sites of production and places of contestation are not arrayed along fixed coordinates but come into contact through practices of folding, stretching, and distorting, playing with the limits of space and spatiality thrown into relief by capital and the state. Social actors increase their affective powers by transecting space, folding in distant harms, and stretching out ties of responsibility to leverage their reach topologically. This impulse is countered by an oppositive drive: powers are decreased by folding out political demands and collapsing affinities to place responsibility beyond reach. Through such manipulations of space, actors effect distortions of distance (proximity/remoteness), scale (local/global), positionality (interiority/exteriority), and prominence (presence/absence). Liberated from the grid of the Euclidean plane, distant harms can be made proximate, just as nearby targets can be placed beyond reach.

Key to Allen's formulation is the dialectical interrelation between labor and capital. Even as labor tries to transect space by folding in, this process is prefigured by capital's oppositive impulse: the folding out of political demands. Successful acts of solidarity make publics present at a site of struggle even when they are scattered around the globe:

> What happens elsewhere, in far-off places, and what is drawn from the past to make the present possible, are all part of the topological equation, where presence does not have to be local, nor part of the same

moment or time period, to be a link in a newly-formed networked arrangement.[26]

But capital counterposes labor's push to stretch ties of responsibility by placing claims out of reach and making proximate struggles seem distant. Thus, if the transnational labor movement, in its best moments, has succeeded in making distant harms proximate, corporate power can displace responsibility, placing nearby targets beyond reach. Distorted through capital's funhouse mirror, what is already quite close may instead appear far. A politics of pity for suffering in faraway places often seemed more palatable than a focus on conditions at home. In topological relations, successful transnationalism may depend less on matching the employer's scalar dimensions on a one-to-one basis than on strategically deploying targeted action at designated points throughout a variegated and maldistributed web of power.

Topology against Networks

Allen explicitly counterposes his power topology to network-based theories. Networks treat distances, scales, and prominence as attributes inherent to network actors rather than meanings constructed through power relationships. Networks theorize connection through the metaphor of bridging (gaps), crossing (borders), or jumping (scale), thereby leaving the territoriality of power unquestioned and reducing solidarity to a question of acrobatics and feats of derring-do. Distance, scale, prominence, relevance, and positionality are not known quantities that can be affixed to some preexisting network schematic but rather are assembled and disassembled in real time through power relationships. In contrast, the topological powers of connection dislocate space through twisting, bending, and other deformations, distorting the apparent fixity of bounded space to traverse it. While a topological assemblage is not unstructured, its structuration is an emergent property that becomes legible through articulations and disarticulations rather than a defining attribute of the system. Authority, influence, and manipulation are practices of power that reshape a commonsensical understanding of what is near and far by "making presence felt."[27] It is a question not of "choosing" which scale to operate on but of acknowledging that many scales already transect the social.

Topology of Automotive Production

The UAW's fitful and halting efforts toward transnationalism in Chattanooga make for a valuable case study. Suffice it to say that a viable transnational labor coalition did not emerge in Chattanooga. Nonetheless, the UAW's experience in Chattanooga provides a unique vantage point through which to examine the pitfalls of ill-conceived transnational projects and the untapped potential of more viable strategies. At some level, the UAW's contest was always already a transnational labor struggle, if only in the minimal sense that an American union was contending with a German-headquartered firm. However, occasional gestures toward solidarity between American workers and their German counterparts proved short-lived, ambivalent, and often self-defeating. Properly contextualized, these problems of transnational coordination in Chattanooga clearly parallel and, in some cases, mirror the labor movement's earlier struggles over globalization, reflecting deeply entrenched political views that mitigate against cross-border solidarity. Moreover, lessons from Chattanooga address gaps in the NGLS literature while suggesting new paths toward a global labor movement.

Reexamining our empirical project through a topological lens provides additional perspective on the challenges of transnational solidarity. VW's labor relations are best understood in topological terms. In 2014, VW headquarters employed techniques of distanciation, including exaggerating the relative independence of its American subsidiary, disavowing responsibility for the behavior of supervisors, and possibly seeding the media with leaked threats to shutter and relocate the plant.[28] But, according to a topological mode, distortions of space are not the exclusive domain of corporations: labor is also equipped to reshape the field of contention and terraform the social landscape to align with its interests.

At VW, folding out had perverse consequences. For all the concern with working conditions in Germany, and for all the focus on importing German-style labor relations to Chattanooga, legacy automakers in the United States were perhaps the more relevant point of reference. The Big Three are just as globalized as their competitors. Without in any way diminishing the potential power of an international alliance between the GWC and the UAW, the *intra*national disconnect between the UAW's northern and southern operations was perhaps a more meaningful point of leverage. In particular, regional pay differentials and six years of postbailout weak contracts remained serious obstacles to any victory in the South. As the UAW now seems to realize, the key to organizing the South may paradoxically lie in shoring up its existing power in the North. A good contract at the Big Three provided powerful free advertising for the UAW. But in the distorted topology of capital, Detroit seemed as far from Chattanooga as Wolfsburg. In the context of VW, folding in would suggest not only a UAW-IGM alliance but also the forging of ties of responsibility between unionized UAW workers and prospective members at transplants by drawing connections across (perceptual) distance.

As in 2014, the works council and the company sought to present German workers as the primary reference group despite the scalar gap and physical removal that separated them from Chattanooga. These workers were drawn in, becoming more prominent, while workers at other UAW sites were folded out. The making-absent of the estimated 30–40 percent of workers working as temps under the subcontracted arrangement is another example of artificial distanciation. Though such workers could not join the union, they were still relevant to the UAW's campaign.

As stated previously, the labor movement leverages topological power by folding in faraway concerns while stretching out solidaristic bonds. In this sense, efforts by the works council to forge ties with the UAW required that the plight of autoworkers in Chattanooga be brought within reach, drawing close what was previously seen as distant. Likewise, a nationally bounded labor movement that traditionally accepted responsibility only for domestic affairs must extend its purview to include the genericized every-worker who knows no country. Allen sees successful transnationalism as bending and reconfiguring space itself, in contrast to the metaphor of "jumping scale," where scale is a fixed hurdle to be surmounted and social actors must train as jumpers. Transnationalism, when it succeeds, is less about making connections across space than about creating new forms of proximity. Capital has already succeeded in collapsing oceans and borders; it is high time that labor did the same.

By 2024, Detroit had replaced Wolfsburg as the locus of the UAW's Chattanooga campaign. The change reflected an informal realignment that impacted meaning making and resituated actors discursively, even without a formal covenant between parties. The UAW folded in workers at Spring Hill and other sites, taking advantage of their (physical) proximity while minimizing the lack of a common employer. During the stand-up strike, the union also resisted distanciation by folding in Big Three–operated distribution centers and making their presence felt through practices of proximity, which led to their inclusion in its master contract. Labor's power depends on the ability to fold in distant harms. For workers in Chattanooga, Big Three contracts did not seem relevant to the 2014 election but became proximate in 2024, exerting intersubjective and relational influence despite having no immediate effect on the election's outcome.

In sum, the enfolding of topological space means that what can be placed beyond reach can also be drawn closer through practices of proximity. Through such topological maneuvers, physical distance can be exaggerated or foreshortened through folding in and pushing away. In sum, power effects distortions of scale, proximity, positionality, and prominence, repositioning relevant actors in the social field and creating the basis for new modes of struggle.

Momentum-Based Organizing

Beyond its topological orientation, the UAW's 2024 victory benefited from a fundamentally different organizing model. The stand-up strike had knock-on ripple effects in Chattanooga, signaling a shift in the organization's priorities and infusing the campaign with vitality. Immediately after its strike, the UAW announced it would parlay its gains into a comprehensive strategy targeting all nonunion automakers, foreign and domestic, EV and ICE. However, rather than take on employers one at a time as it had previously, it declared its intention to organize all transplants simultaneously. Confronting them all at once created a sense of movement across states and regions rather than the geographic isolation of one-off campaigns. Fresh from its groundbreaking stand-up strike, the UAW returned to Chattanooga in 2024 with a renewed fighting spirit.

As long as the UAW kept winning, the simultaneous approach would allow the union's victories to snowball and fuel concurrent struggles. Rapid-fire organizing drives create hot shops[29] primed to organize and feed off previous victories. Building velocity is key to any growth strategy because historically, when the labor movement has added members, it has not done so steadily or predictably but rather in fits and starts. Moments of growth tend to be sudden and dramatic, while periods of stagnation and decline are often more prolonged and gradual.[30] While it achieved some significant victories at small factories in the 1990s and 2000s, these were not high profile and tended to be overshadowed by larger losses. For the first time in decades, the UAW was poised to ride the accelerating tide of momentum. *Labor Notes* described the regional organizing drive as a "relay race," succinctly capturing its dispersed, multisite character.

While past performance is never a predictor of future results, prospective members evaluate a labor organization primarily based on its recent accomplishments. In 2024, the UAW's sales pitch to members included its sweeping victories at the Big Three, which served as a source of pride, not embarrassment. By 2024, the union had moved from a defensive to an offensive position.

Newcomers to the labor movement are often reminded that organizing is a laborious chore requiring meticulous attention to detail, obsessive "workplace maps" that leave nothing to chance, and systematic assessments of each worker to evaluate support before going public. Jane McAlevey's aptly titled *No Shortcuts*, which has become something of a movement bible for a new generation of union activists, is perhaps the strongest expression of this perspective.[31] The book is both a how-to manual for organizers and an unsubtle critique of organizing styles that seek to bypass the methodical approach she defends. But McAlevey's lessons are drawn from her time with HERE during the 1990s, a challenging period for labor when victories were few and far between. McAlevey was, by all accounts, an enormously talented organizer who repeatedly defeated A-list employers. However, her carefully orchestrated wins could not generate the "contagion effects" associated with social movements as they did not occur in the context of a labor upsurge.

To be sure, even during the 1990s, momentum-based organizing had its advocates. But these voices were confined to the fringes and often dismissed as naive, irresponsible, or adventurist. Momentum cannot be willed into existence; it requires a confluence of factors that were hard to imagine at the time. Momentum organizing begins with so-called trigger events, which initiate a chain of events that quickly spiral out of control. According to Mark and Paul Engler, the authors most closely identified with momentum organizing:

> Trigger events are highly publicized moments that mobilize people outside of existing structures. . . . If the trigger event is big enough, then there is an outpouring of decentralized energy that emerges.[32]

In the context of the UAW, the trigger event was the stand-up strike, which received enormous attention and largely positive coverage. By the time it reached a contract, polling showed 87 percent of Americans were aware of the strike[33]—a remarkable achievement considering most news desks no longer hire labor reporters. Chris Brooks, who would later take a position inside the UAW and become the primary in-house proponent of momentum organizing, writes:

> The goal is to foster a virtuous cycle of building to trigger events and then absorbing the subsequent explosion of energy through mass trainings and decentralized structures, while then building to another, future, trigger event. When the whirlwind comes, what was once seen as a risky long-shot action or fringe idea suddenly snowballs into a series of independent, self-organized actions.[34]

It is also easy to understand why advocates of structure-based organizing might scoff at the momentum approach.[35] Structure-based organizing errs on the side of caution, delaying election filing until 70–80 percent of workers have signed union cards. In contrast, momentum organizing is definitionally risky, verging on reckless. Momentum-based organizing advocates have little interest in the painstakingly updated spreadsheets and scripted one-on-one "organizing conversations" that have long been a mainstay of traditional approaches, speaking instead about "explosions of energy" and "whirlwinds." For those steeped in the McAlevey school of organizing, the momentum approach seems about as relevant to their method as alchemy is to modern chemistry.

Yet momentum-based organizing has real dangers. The same forces that allow a union drive to catapult itself forward at record speed may leave it vulnerable to charismatic leaders who whip up frenzy without laying the groundwork to sustain the organization. Even momentum-based organizing's leading evangelists concede, "In most conditions, momentum organizing is not the way to organize unions."[36]

Therefore, to be precise, the UAW's strategy at VW can best be described as a mixed-method approach, combining momentum with tried-and-true elements of structure-based organizing. Organizers worked on an accelerated timeline, playing up emotional and affective ties to amplify ambient sentiments that already pointed in their favor. Then, just to be sure they had not misread the moment, they filled out their spreadsheets anyway.

Toward Sectoral Bargaining?

Combining a topological orientation with momentum-based organizing has positioned the UAW within striking distance of the labor movement's holy grail: sectoral bargaining. Despite domestic automakers' close association with Detroit, they have maintained assembly well south of the Mason-Dixon Line nearly since their inception. Geographically, there has never been a neat separation between North and South. Indeed, two domestic plants are more physically proximate to Chattanooga than the closest transplant. Previously, the UAW had neglected to make these connections explicit, erecting strategic firewalls between the organizing targets that severed its "southern strategy" (such as it was) from the arm of its organization that handled domestic employers. Although the UAW paid lip service to "organizing the South," harking back to its failed Operation Dixie, its push to organize the South amounted to a series of one-off campaigns, as scattered geographically as they were isolated strategically. A recurring pattern began to emerge: roughly every two to three years, the UAW would either file for an election at a foreign-owned automaker, only to cancel the vote as the opposition ramped up and support dwindled, or else suffer a decisive defeat. This one-at-a-time approach led to a focus on the specificity of particular automakers rather than recognizing their commonalities. As noted in Chapter 2, in every case where it had gone up against a foreign automaker, the UAW faced stiff resistance, union busting, an organized anti-union lobbying effort from community groups, and, in at least a couple of cases, betrayals and dashed promises. To be sure, there are important differences between the three German automakers, not to mention between German firms and their Japanese counterparts. However, the narcissism of small differences blinded the UAW to the similarities that persist among the thirteen foreign auto companies now producing vehicles in the United States, which together own thirty-six plants that employ approximately 150,000 workers.

As noted in Chapter 4, a "global" campaign based on a single company's transnational footprint will inevitably lead to company-specific exclusions. For VW, until recently, the United States was a minor link in its global production network, producing fewer than 1 percent of its cars, compared to 15 percent of total automobile production across all companies. By focusing exclusively on VW (unionism in one company), the UAW confined its campaign to a site with little strategic value from the company's perspective. By recontextualizing the arena of combat, the UAW made visible linkages between VW and other German firms while universalizing what had until then been a bespoke campaign centered on a statistical outlier.

Conclusion

In their boldest moments, these developments point to a new modality of global solidarity that surpasses now-discredited forms of transnationalism. Labor might do well to escape the territorial limits that separate actors from benefactors and turn instead to what might be termed *infranationalism*. This concept borrows from Robin D. G. Kelley's use of *infrapolitics* (originating in James Scott's work on Southeast Asian stateless people) to describe daily acts of resistance and survival that remain invisible to power, never rising to the plane of "traditional" politics.[37] If an explicit engagement with states and power structures characterizes "traditional" politics, infrapolitics is characterized by a subversive undercurrent of everyday conflict that never registers as a form of political engagement. Similarly, infranationalism suggests an evasive subjectivity that avoids codification through the statist apparatus.

Recent events have led to an epochal shift in the automotive industry, repositioning actors across the social field, linking them not by shared productive capacity but through discursive frames. The 2023 strike set in motion a discursive shift, leveling the playing field so neither region nor country of origin is decisive in predicting company behavior. Rather than racing to the bottom to match the low bar set by foreign competition, the Big Three have now established a ceiling for foreign automakers, who are incentivized to catch up to remain competitive. Simultaneously, a newly activated UAW has proved capable of contesting elections in the North and the South. Again, this transformation was both material and discursive, encouraging a perspectival reorientation toward place and space. In short, foreign and domestic firms were interwoven, creating a latticed structure that empowered all workers irrespective of their firm's nationality.

CONCLUSION

Solidaristic Futures?

This book has adopted the strategic capacity framework as its primary analytical lens. However, I departed from Ganz's original formulation in one key respect. Ganz treats heuristic processes as a single analytical category, combining recursive learning with speculative, innovative, and creative processes. Learning and innovation may, at times, be complementary and mutually reinforcing, but their covariance cannot be assumed. Distinguishing between learning and innovation facilitated a comparison of the UAW's 2019 and 2024 campaigns. Innovative capacity increased dramatically over this five-year cycle, while learning remained more consistent. We concluded, therefore, that learning alone cannot build strategic capacity because even the most acquisitive "learning organization" is hobbled by deep structures that preserve the status quo and hamper true innovation. Until 2023, the UAW remained ill equipped to resolve the situation without deep structural change, or what we have called innovation.

A single case study lacks external validity and is insufficient to draw broader theoretical conclusions. Still, this study points to the need for further research to validate the heuristic processes construct and clarify the relationship between learning and innovation. The strategic capacity model's great strength and fundamental weakness is that it is relatively self-contained. As noted previously, Ganz's model is built on the example of United Farm Workers, which operated within an industry that had been purposely excluded from the NLRA and that thus was less dependent on the state. Moreover, growers exerted oligopolistic control in California, but they were few in number and nationally based, and their work was confined to the national market (with some overlap in Mexico). While national capital is an oxymoron, the terrain of struggle was relatively bounded. Expanding Ganz's ideas to the UAW necessarily exposes them to certain problems. When dealing with more complex organizations operating on a variegated terrain, accurately measuring "strategic capacity" may be less straightforward. As the aperture widens or narrows, it becomes impossible to know in advance which information might be relevant, which actors are to be motivated, or which lessons will prove instructional for learning.

This chapter projects what we have learned about strategic capacity into the future, with attention to the emergent EV sector. As noted in Chapter 6, EVs became a secret weapon for the UAW. Few would have expected that the EV revolution would work in the union's favor, but EVs helped mitigate the threat of relocation in 2024, reducing what might have otherwise been a serious liability for the union. However, going forward, as EV-related infrastructure becomes more prevalent and companies build up redundancies, VW workers will be unable to count on EV quotas to save the day.

Vulnerabilities

We open with a discussion of supply chain vulnerabilities and logistical choke points. Despite some openings, the UAW's strategy did not avail itself of structural power, with the partial exception of 2024.

The Chattanooga Assembly Plant is deeply embedded within a dense network of codependent suppliers. The plant was considered state of the art when built, with numerous features representing the latest thinking in industrial design and supply chain integration. True to this vision, the three major production areas—the body shop, assembly, and paint shop—were closely integrated and situated in a concentric ring to minimize the distance between stations. In keeping with German production models, in which companies maintain close relationships with suppliers who often enjoy exclusive contracts, VW built an on-site supplier park with sixteen key component manufacturers in mind.[1]

Beyond this immediate base of operations, the plant aimed to source 80 percent of its supplies locally—an ambitious goal that would not immediately be achieved, given the limited production capabilities of the Greater Chattanooga metropolitan statistical area (encompassing a ten-county region in southeastern Tennessee and North Georgia).[2] Seven first-tier parts suppliers in an on-site business park produce crucial components and provide 1,100 jobs, in addition to the roughly 5,500 in the main factory proper.[3] A stamping plant just miles away accounts for 500 jobs, and an axle assembly plant, another 240 jobs.[4] No longer content to source its battery cells from Asia, VW recruited a lithium hydroxide manufacturer (117 jobs) and a synthetic graphite plant (300 jobs).[5]

Thus far, the UAW has yet to organize many of these suppliers. Though the UAW made some slow and steady gains in the 2010s among smaller parts distributors, these victories have not yet reached snowballing velocity. Yet, as noted in Chapter 7, successful organizing efforts can provide others with a road map and lead to copycat actions, which eventually create contagious waves of protest through their "demonstration effects."[6] There is some evidence that this occurred after the win in Chattanooga. Region 8 organizing director Tim Smith told the local paper of record that he had received phone calls from suppliers across the South: "I can't tell you how many calls come in daily from workers at these (parts supplier) plants."[7]

An account of transnationalism that takes the topology of power as its starting point rather than the bounded geography of the firm would surely acknowledge the centrality of these subcontractors, whether or not they hang the VW logo over their marquee. Indeed, the UAW has already organized seating plants and a stamping plant in the region.[8] VW's supply lines are so tightly integrated that a strike or slowdown at any of these suppliers could easily bring production at VW to a halt, putting pressure on the company to recognize the union.

Under just-in-time (JIT) production, parts are made available on an as-needed basis with short turnaround windows to reduce the need for storing excess inventory and guard against overproduction. Under ideal conditions, JIT enables firms to closely monitor their suppliers' output, place orders on short notice, and resolve any issues in real time before they create backlogs.[9] However, while JIT creates efficiencies, it also introduces new forms of risk, as taut supply chains lack agility and resilience if a disruption occurs.[10] Minor disruptions at any level of the process can have ripple effects that eventually cause turmoil across the entire system and jeopardize the timely delivery of finished product.[11] Even worse, low-impact disruptions might affect multiple critical nodes simultaneously, thus magnifying their impact and increasing the overall vulnerability of the entire supply chain.[12] Moreover, modern supply chains are so deeply integrated that disruption by a mid-tier supplier can have both upstream and downstream effects, potentially affecting both lower-level subsidiaries and final assembly.[13] In the worst-case scenario, a disruption within a colocated cluster could have a domino effect, leading to the toppling of the entire cluster.

Labor law restricts the ability of unions to exert pressure via supply chains, but unions have options even within the constraints of the law. So tightly wound are these supply chains that the mere threat of a strike can have a serious impact, at least at carefully selected choke points in the global movement of goods and capital.[14] Given the competitive pressures mentioned previously, suppliers are especially concerned about disruptive employee behavior since suppliers with reliability issues may be refused future contracts. Thus, job actions by workers at these firms may exert disproportionate power, wielding influence vastly exceeding the direct revenue loss they inflict on their employer.[15]

Labor has occasionally taken advantage of capital's self-inflicted supply chain vulnerabilities. In 2014, striking workers at a small and relatively inconsequential supplier gained full recognition within a matter of hours due entirely to the fact that this firm happened to be the sole source of brakes and struts for Chrysler's Jeep plant on the other side of town.[16] Even less common is the use of job actions at a subsidiary to force concessions on a purchaser, though such a scenario is certainly not beyond the realm of possibility. Even barring job actions, JIT supply chains in their pure form do not accommodate contingencies or allow for redundancies, as such precautions would require major sunk costs. Occasionally, the weakness of this model is sharply revealed, as it was when a fire at a single auto parts supplier that happened to figure prominently in Ford's JIT supply chain forced the closure of three assembly plants in two states until the parts could be ordered and delivered from an alternative.[17] During the stand-up strike, choke point vulnerabilities prompted Stellantis to shift inventory out of its aftermarket parts distribution centers in preparation for a strike that never came.[18] Similarly, the mere threat of a strike at Allison Transmission in February 2024 led to the elimination of tiers at the supplier plant.[19] As Kim Moody notes, strikes targeting suppliers and logistics hubs might circumvent prohibitions on secondary boycotts and sympathy strikes, as production impacts up and down the supply chain can be realized without the "deliberate" targeting that runs afoul of the law.[20]

For all these reasons, a countercurrent has developed within logistics studies that questions the received wisdom of colocation. Heterodox scholars have come to question the effects of clustering, noting that, despite the important advantages of supplier concentration, certain unintended consequences may arise, including increased vulnerability to a single disruptive event.[21] The negative downside of clustering was powerfully revealed by the devastation of Japanese production in the aftermath of the 2011 tsunami—an impact that was worsened because Toyota and Honda colocated the vast majority of their suppliers in physically proximate supplier parks.[22] For this reason, dispersing suppliers across a broad geographic area may be a preferred sourcing strategy despite longer lead times, increased opportunities for transportation bottlenecks, and less intersupplier collaboration.[23] In an empirical study of supplier firms, five of the seven firms noted that geographically clustered suppliers increase the likelihood of disruptions within a supply chain.[24] Drawing on another study, scholars found that supplier "density" (clustering) is, in fact, a liability, not an asset, in the event of a disruption: "An unplanned event that disrupts a dense supply chain would be more likely to be severe than the same supply chain disruption occurring within a relatively less dense supply chain."[25] In this view, clustering decreases resiliency and increases vulnerability.[26] At a minimum, there is growing recognition that both colocation and dispersal have benefits and drawbacks. Striking the appropriate balance involves the messy practice of balancing risks and rewards.[27]

Importantly, as firms grow their value chains, there is a pressing need to mitigate risk and minimize disruption. Together with JIT, the geographic concentration of production by the Big Three in the 1930s created multiple choke points, allowing small groups of autoworkers to exert undue influence through the strategic use of the strike weapon. Supply chain management rarely makes headlines until something goes wrong, but the rigidity of the supply chain was revealed to the general public by enormous disruptions during the COVID-19 pandemic and the threatened 2022 rail strike.

Additional vulnerabilities may result from the intersupplier competition inherent to the cluster model. Although clusters are premised on manufacturers' monopsonistic purchasing power, it is in their interest to compel suppliers, transporters, warehousers, and other downstream firms to compete against one another for contracts. Colocated suppliers are selected only after entering a competitive bidding process and must continually minimize costs under constant threat of replacement. Again, this reality has important implications for VW. By its own admission, the company was drawn to Chattanooga because—unlike most similar-sized metropolitan regions—it is dual served, meaning multiple railroads (CSX and Norfolk Southern) can access the facility. In practice, there is strong evidence that VW has played the railroads against each other, influencing CSX to upgrade its intermodal facility in an effort to secure VW's business. Yet, in the case of CSX, this has created yet another choke point in the form of an intermodal transfer hub just across the Georgia line from Chattanooga, which handles the vast majority of all Savannah-bound cargo on the rail line.

A strategic understanding of these choke points and vulnerabilities can only benefit the UAW as it looks to defend and expand its victory in Chattanooga amid the rise of EVs.

Electrification

As noted in Chapter 7, Chattanooga's pivot to EV production dramatically benefited the UAW in 2024, strengthening its hand and guarding against the threat of disinvestment. But examined in its totality, the EV transition's impact on VW, the UAW, and the industry more broadly remains unclear. This section explores the implications of the rise of EVs on autoworker power, both as applied to VW in Chattanooga and in a more general sense.

As this book went to press, several important battery-related developments occurred that had implications for Chattanooga. First, it was announced that Chattanooga would terminate its agreement with South Korean–headquartered SK Battery and begin insourcing its battery production through a new wholly owned subsidiary, PowerCo, which will open its first North American plant in Ontario, Canada.[28] Second, VW announced plans to revive the Scout brand as a 100 percent electric SUV label at a site in South Carolina. Whether the plant will allow card check is unclear, but this will be key as the UAW seeks to extend its Chattanooga victory.[29] Finally, VW entered into a joint venture with the beleaguered EV truck company Rivian.[30] Together, these developments signal the growing uncertainty that marks the transition to EVs.

The initial disruptive force and context for the restructuring of the automotive industry was the automobile's deliberate progression toward electrification, coupled with the eventual obsolescence of ICEs. Battery plants pose a major challenge to the automotive global production network, as laid out in Chapter 1. By the 1980s, the geographic coordinates and hierarchical logic of automobile production were unsettled through vertical and horizontal disintegration. They now face further disruption as the EV transition restructures the industry, unbalancing global logistics and elevating some actors while demoting others.

Geographical Displacement

The unique features of EV production have hastened a spatial reorganization of the automotive industry, challenging time-honored customs and calling into question long-standing assumptions. The clustering of suppliers and final assembly is an established trend, with companies seeking to minimize the costs and risks associated with long-distance supply and transport to market. In some ways, EVs continue this trend, with many first-tier suppliers of sensitive and critical components—like notoriously volatile lithium batteries—colocating near host companies. However, new considerations have disrupted the logic of clustering. The EV industry has altered the calculus since automakers may now value proximity to resources.

The spatial distribution of EV uptake does not mirror ICE consumption patterns, with select urban coastal communities far outpacing interior and nonurban areas. (This patterning is driven by various factors, including southern leaders' reluctance to fund EV infrastructure such as fast charging stations, the high cost of EVs, and the ideological impact of climate change skepticism.) In keeping with these trends, VW's newest battery plant will be in Ontario, across an international border and 750 miles from its nearest production plant. The choice was made in part because of proximity to lithium reserves—which are abundant in Canada but more scarce in the United States—as well as the country's proactive approach to environmental social governance, a must in an industry that prides itself on green credentials.[31] The new facility will produce solid-state batteries, which are less volatile than lithium, but production is not expected to begin until 2027.[32]

Geopolitical Realignment

Geopolitical concerns also drive the regional patterning of EV production. Unlike other aspects of automotive production, batteries remain heavily dependent on foreign suppliers. The Biden administration's cool reception of "foreign" EV manufacturers and their first-tier suppliers can only be understood through a geopolitical lens. The push to incentivize domestic EV production may be sound economic policy, but it must also be understood as the opening salvo in what will likely be a prolonged international conflict with profound ideological, political, and perhaps even military implications. The United States' continuing dependence on the foreign superconductor industry is a source of national embarrassment, not only because lengthy supply chains are vulnerable to disruption but because the country's entanglements with its rivals compromise its position on the world stage. Secondarily, the dramatic turn toward EVs stems partly from a growing uneasiness with the petrocapitalist states that underwrite ICE production. Tariffs and import substitution may aim to incubate a nascent domestic EV supply chain but serve a dual function as state propaganda.

The disconnect between industrial policy and geopolitical concerns came into stark relief in the Inflation Reduction Act as initially drafted, which established incentives for domestic sourcing whose baseline qualification threshold would be unachievable for any American company in the near term. At the time of passage, the American battery industry could not produce batteries at the rate necessary to qualify carmakers for subsidies. In its quixotic campaign to encourage the growth of the American EV industry, the Inflation Reduction Act was defeated by its hubris. In this case, ambivalence in global production manifested in the form of a policy that, through its very ambition, strained the limits of possibility.

Although VW pledged to onshore production and remove any Chinese-made components from its vehicles, the reality of sourcing superconductors, batteries, and the metals required to produce batteries has left VW and other firms heavily dependent on China.[33] Chinese-based suppliers are actively marketing themselves to VW in open defiance of the company's pledge to cut China out of its supply chain.[34] Like other OEMs, VW, no longer content to allow component suppliers to manage their own sourcing, is now striving to exert tighter control over its supply chain.[35] Even as it ramps up domestic production and invests heavily in downstream North American EV parts, it has struggled to avoid sourcing Chinese subcomponents, which, on at least one occasion, caused shipping delays when contraband parts were uncovered.[36] Thus, VW finds itself in the paradoxical position of making battery parts close to home while also extending supply lines to China.

Horizontal Disintegration Redux

GM took a new tack at its battery plants, creating a brand-new firm, dubbed GM Subsystems (GMS), which it owned and managed, to oversee a substratum of workers otherwise undifferentiable from direct hires. Though temporary and subcontracted workers nominally worked under different contracts or were supervised by a different firm, the difference between GM Subsystem and GM proper is purely notional. GMS was founded as a wholly owned subsidiary that initially operated only in battery plants and paid barely 60 percent of the starting wage stipulated under the master agreement. A major departure from past practice, GMS has been described as an "invented category" or "a legal shell to evade its main union contracts."[37] Understood as horizontal disintegration, GMS differs from traditional subcontractors in that the work performed is not of a lower order or lesser skill grade than core proficiencies but is identical in all but name. (GMS workers are generally assigned to warehousing and material-handling tasks, which are considered core competencies at OEMs.) At equivalent ICE plants, no such entity exists. Pioneered at GM's plant in Brownstown, Michigan, a similar pseudosubsidiary model was quickly adopted at Ford for its co-owned battery plant in Marshall, Michigan.[38] Inside the plant, GMS workers are indistinguishable from their directly employed counterparts, sharing the same spaces and performing the same tasks (except for jobs requiring special training or technical expertise). Thus, even if a strict demarcation between temporary and permanent employees retains its legal basis, it fails to capture the authentic experience of the division of labor on the shop floor.

During the stand-up strike, the UAW made bold strides in connecting GMS and other subsidiaries to its legacy ICE contracts. GM is wholly dependent on its GMS workers, many of whom are strategically located near its final assembly plants. Indeed, GMS is a core competency in all but name, serving as the company's exclusive favored supplier and barred from bidding on competitor contracts. The fact that these workers are not counted among GM's core workforce is a mere accident of corporate accounting policies. An analysis of production that takes as its starting point the topology of power rather than the bounded geography of the firm would surely acknowledge the centrality of this subsidiary, whether or not it hangs the GM logo over its marquee.

Topologies of Electric Power

The emergence of EVs is best understood in topological terms. While it may be tempting to fold out EV suppliers and view them as an unwelcome incursion and threat to ICE jobs, this will have perverse consequences given the relational ties between EVs and ICEs. The perception of distance may be unrelated to physical proximity. The forging of ties of responsibility between unionized UAW workers and prospective members at transplants and EVs draws connections across (perceptual) distance.

The status of battery and battery-component plants in relation to final assembly workers remains fiercely contested. Battery plants are paradoxically situated in relation to the industry more broadly. On the one hand, they are a "technological bottleneck" or strategic choke point on which the rest of the sector depends.[39] Therefore, the success of any green transition (to which the UAW is nominally committed) depends, in large part, on the success of these firms. On the other hand, EV production is plagued by lower wages, worse working conditions, and more pervasive health and safety issues than ICE production.

As noted in Chapter 7, while the UAW initially viewed EVs as an unwelcome incursion into its established strength in ICEs, the union has more recently attempted to leverage industrial policy to strengthen its hand. Initially, labor raised some forceful objections. Until 2023, battery plants owned by or affiliated with Detroit's unionized Big Three often operated under separate agreements with lower pay and fewer fringe benefits. Therefore, the UAW initially responded to EVs with open hostility, attacking a $9.2 billion loan made to Ford for a joint venture battery plant that was to operate outside its existing agreement.[40]

But under Fain's leadership, it aligned (if somewhat reluctantly) with the Biden administration's push for electrification, advocating for stronger worker protection under the Inflation Reduction Act.[41] This is an exemplary case of labor cooperating with the state to drive innovation and incubate a nascent industry. After initially adopting an adversarial stance toward outsourced work, labor has seemingly reconciled itself to the inevitability of change.[42] Even initially, the UAW's qualms over upgrading often expressed themselves, not as outright resistance, but as ambivalence, or what Adrien Thomas and Nadja Doerflinger describe as "hedging"—a mealymouthed phrase that concedes that the battle over technology has already been lost.[43] As Mathieu Dupuis et. al write:

> The acceptance of change may be because of the tightness of the labor
> market; because the amount of future job loss is uncertain; or because

of the widespread view that new investment is needed for the long-term viability of unionized factories.[44]

Such debates over the "just transition" reflect concerns over the relevance and proximity of EVs to ICE production. EVs' positionality relative to ICEs remains uncertain. However, the UAW has made bold strides in connecting the EV sector to its legacy ICE contracts.

In a departure from its protectionist past,[45] the UAW prioritized the achievement of full parity at the battery plants during its 2023 strike and eventually succeeded in incorporating Big Three–affiliated battery plants under its 2023 master agreement.[46] While this development somewhat allayed concerns over electrification, there are still stark divisions between battery workers and conventionally employed autoworkers.[47] Still, the UAW is increasingly adamant in its insistence that plants be unionized and that wage and benefit guarantees be comparable to those of ICE plants. Improving the status of EV-related jobs is a high priority for the UAW and was among the main factors motivating its 2023 strike. The status of battery plants is murky under U.S. labor law, as expansion of the bargaining unit is not covered among mandatory bargaining subjects.[48] Nonetheless, the UAW was able to insist that they be included in the contract through a concerted campaign of public pressure.

Through the topological manipulation of space, actors redefine who matters and what counts, bending and twisting affiliatory ties. Though downstream from the lead firm, battery plants are arguably the primary drivers of outcomes because they are a choke point for a precious resource and because they transect firms, nations, and regions. Distinctions between home and host countries are not meaningful when applied to battery plants, and codependencies position battery production both *within* and *beyond* the known parameters of existing chains and networks. The EV and battery industry, led by Tesla, has worked to present itself as wholly distinct from the existing domestic and foreign EV and ICE markets. Discursively, capital has placed EVs and batteries beyond reach. On the other hand, labor has repeatedly sought to stretch ties of responsibility to incorporate EVs, first by attempting to add them to the Green New Deal and then by integrating them under its contract.

Applied to the automobile industry, the deployment of new narrative frames is realigning the discursive coordinates of the social field.[49] In 2023, the UAW made important strides and broke with past practice in folding in Big Three–operated EVs by making their presence felt through practices of proximity, which led to their inclusion in its master contract. As tremendous as these decisions were, sites of unevenness and "dark places"[50] considered outside the purview of the UAW remain and are likely to persist. First, the Big Three have followed the lead of the transplants in aggressively hiring temps, who now constitute as much as 20 percent of the overall workforce and are only partially covered under the 2024 contracts. Second, the two-tier system, a by-product of the 2008 recession, was weakened but not dismantled. Third, EVs and related parts are manufactured through joint ventures between the Big Three and foreign companies. The UAW did not contest such joint arrangements during its strike.

This section has shown that EVs became relevant not simply because of their strategic position in a production network but also because they were discursively positioned as relevant actors. Faced with the onslaught of third-party battery and battery-component suppliers, the UAW could have easily cast the emergent sector aside or battened down the hatches in a neoprotectionist maneuver echoing its long history of nationalism verging on xenophobia. Instead, by incorporating foreign battery makers under its ambitious organizing plan, the UAW has signaled that it views batteries not as an outside threat to its "base" (imagined as unionized American workers) but as a rising stratum of a shifting industry for which it bears full responsibility. In doing so, it signals to its members that batteries are not a danger to be combated but an opportunity to extend the UAW's organizational reach. The UAW's success in organizing VW's EV suppliers will depend in part on its ability to resist efforts aimed at distanciation and instead extend its reach to include the nonunion EV sector.

Reinventing Solidarity

In the final section of this chapter, we return to the question of solidarity introduced in the Introduction and elaborated in subsequent chapters. How did the UAW use its strategic capacity to build solidarity in 2024 but not in 2014 or 2019?

Solidarity has a special place in labor's cultural canon; it is the stuff of songs and legend. For Richard Hyman, solidarity can only ever be a Sorelian myth, an asymptotic ideal that is "unrealizable yet perhaps capable of inspiring action which results in its partial accomplishment."[51] But if solidarity has assumed something of a mythical status, it remains misunderstood. Experience reveals just how imperfect attempts at solidarity can be. In a certain vulgar imaginary, solidarity of nearly any variety is inherently emancipatory, helping labor escape from its place-bound fetters and advance toward its internationalist mandate.

Against this simplistic (if appealing) story, Rebecca Johns contends that solidaristic practices can be extensions of business unionism that accommodate rather than transform social relations.[52] Conventional models of solidarity that center on the final product risk a false front that is not backed by meaningful commitments. Though she rejects the simplistic framing of "solidarity from above" and "solidarity from below," one could think of "deep" versus "surface" solidarities.

Johns expounds on the prospect of a "dark side" of solidarity by presenting an accommodationist model in which spatial interests dominate class interests, which she positions against "transformative solidarities," in which universal class interests trump spatially derived and rooted interests. In its worst forms, accommodationist solidarity conceals an antiliberationist project that benefits a specific group of workers over and against others. Such "solidarity" movements come to function as a mirror of reactionary protectionism. Thus, Pat Buchanan finds his perverse appositive pairing in Bernie Sanders. In contrast, the transformative model seeks to equalize both social conditions and investment in job creation instead of prioritizing one set of workers over another. Transformative solidarity "prevents capital's use of the development seesaw to weaken class organization."[53]

If there is a weakness in Johns's logic, her framing suffers from a class-space binary that is perhaps all too convenient. *Space* serves as a stand-in for capital, whereas *class* is a proxy for labor. She risks presenting space as a fixed obstacle to overcome, ignoring that class conflict often reorganizes and reconfigures the spatial arrangements that once seemed to constrain it, just as the class is recomposed through the very process of solidarity building. Complicating matters further, class relations are themselves refracted through their spatial dimensions, making it impossible to neatly separate class from space. Moreover, Johns's vision presents a romantic version of solidarity that relies on selfless altruism.

Johns's rejection of accommodationist solidarity is well taken, but "transformative solidarity" is an unsatisfying alternative to the abovementioned reasons. Solidarity bridges difference not by positing a false equivalence or a communion of equals but by straddling an aporia that cannot be reconciled. Solidarity implies both difference and connection, not the abrogation of one by the other.[54]

First, solidarity implies not universalism but an alliance that transects difference. In social theory, many versions of solidarity are premised on a humanist ideal that precedes—and ultimately transcends—conflictual social relations. Thus, even when social actors become locked in conflict or seek to actively undermine one another's interests, objective shared interests ultimately connect the warring parties. Emile Durkheim, for example, proposed that "a certain number of states of consciousness are common to all members of the same society."[55] Likewise, Antonio Gramsci saw solidarity as breaking down divisions of skill and craft, noting that the peasantry's solidarity would reconfigure "craft particularisms."[56] For Gramsci, solidarity is a technique for sublating petty differences. Doug McAdam, Sidney Tarrow, and Charles Tilly offer a more recent example of universalist solidarity, arguing that solidarity must be based on emulation or likeness: "Successful brokerage promotes attribution of similarity, while unsuccessful brokerage promotes the recognition of difference."[57] But with difference subsumed, the "unmarked" subject becomes the universal subject. Instead, as Sebastian Garbe suggests, presenting solidarity in the first instance as a "conflictive relationship"[58] allows a consideration of how those conflicts are worked through, which might include questions of agency, the reproduction of racialized structures, and other challenges. The key to solidarity is a connection that transects difference, preserving the essential dissimilarity between parties rather than erasing or transcending it.

Relatedly, we must guard against a romantic and naive conception of solidarity that abolishes difference under the banner of working-class unity. The working class may have no country, but a sizable slice of the activity occurring in its name has had a baldly protectionist character.[59] As political theorists have observed, coalition-based alliances are not a "meeting of equals" based on shared qualities or some essential unity of man but a coming-together across difference rooted in inequality.[60] As a corollary, the particularisms that threaten to split alliances should be seen not as a failure of coalitional politics to overcome but as the real and natural expression of the tensions on which the coalition is based.

Second, solidarity is averse to essentialist containers. Rather than uniting fixed and stable groups (nation, ethnic identification), it is a process of social decomposition and recomposition along solidaristic lines. In forming solidaristic alliances, actors reshape their very identities, emerging differently from how they began. Many versions of solidarity view their project as little more than the amalgamation of otherwise distinct national campaigns. But solidarity troubles fixed concepts of identity and displaces social actors through its operations. This notion of solidarity is grounded in relationships instead of identities; it is an effect of social relations, not their cause. Solidarity might be conceived not as an alliance or a coalition, which presumes a certain fixity on the part of participants, but through an assemblage model, which escapes the hierarchical prison of conventional models, for it is a "generative interaction, which can be neither reduced to its parts nor expanded to an infinite totality."[61] It follows that, just as solidaristic actors are malleable, so is the context in which solidarity takes place. In this sense, the "normative" dimension of solidarity is activated and comes to define reality itself through its dialectical coarticulation.

Third, while theorists of solidarity focus on its "positive" or generative aspect—unification across difference—solidarity also has an antagonistic component: shared commitment against an opponent. Solidarity arises in opposition to an outsider who is not party to the solidaristic relationship but is nonetheless implicated in the arrangement, constituting what Jodi Dean calls "a third."[62] The connective tissue of solidarity forms an alliance that redirects itself toward the "third." Its success depends as much on the compatibility of would-be allies as on a shared orientation toward this "third." Thus, not only does solidarity involve the intersubjective negotiation of the solidaristic bond; it also requires coordinating a public-facing strategy that projects this bond outward toward a generalized Other. The "third" is no neutral observer but a self-interested party who actively intervenes, disrupting solidarity whenever possible through divide-and-conquer countertactics. This three-way set of relationships, not any subset thereof, is the authentic basis for solidarity. Just as important, solidarity imagines itself being watched, taking on performative attributes, and putting these relationships on display. Likewise, solidarity statements describe a social situation of shared interests while shaping that very situation through solidarity's performative display.

Fourth, solidarity's strength and durability depend on its composition. Authentic solidarity cannot be built on a charity or (neo)colonial model. Research on solidarity must be attuned to "the labor of assembling and making connections."[63] If solidarity is treated as a process, not a principle, identifying those responsible for constructing its conceptual architecture becomes crucial for understanding its machinations and exclusions. Exposing solidarity's foundations also allows for contesting the terms on which relations are generated. If solidarity is contingent and ongoing, it follows that it can be redirected or rebalanced rather than simply supported or opposed as a matter of principle. Accommodationist solidarity ultimately favors the gracious benefactor who bestows solidarity on passive recipients, incapable as they are of fully articulating their interests, much less reciprocating. In this gesture, solidarity is produced at the highest level and percolates downward along a tiered system of power relations. If we are not attuned to these imbalances, what appears to be solidarity risks "the inclusion of knowledge production on the one side and exclusion of the local translators and originators of these debates on the other."[64] The beneficiaries of solidarity, which is indistinguishable from charity, can only express their gratitude, as they lack the wherewithal to understand their needs.

Fifth, solidarity is an intensive endeavor that the architects of global governance cannot reverse engineer. Solidarity has been hamstrung by what Richard Hyman has called "bureaucratic trade union internationalism"—an orientation resting on the tripod of economism, the nation-state, and technocracy. Indeed, based on the sheer extent of the bureaucratic apparatus ostensibly dedicated to transnational solidarity, one might expect a higher success rate. The solidaristic project suffers not from a lack of organizational capacity but from the failure of a solidaristic imaginary. In contrast to cookie-cutter and predictable solidarity statements churned out as a matter of principle, solidarity is spontaneous, inventive, and creative. Under this model, solidarity is contested and subject to negotiation, deriving its meaning not from abstract principles but from congruencies arising through shared struggle.

Sixth, solidarity is an open-ended relationship without guarantees rather than a unidirectional march toward alliance building. Featherstone suggests that solidarity is nonteleological, having no a priori signification, and "must be constructed politically," and it is therefore contingent on its activation through struggle. This model of solidarity does not merely incorporate difference as an adjunct to universalism but foregrounds difference, which must then be worked through to unearth commonalities: "These processes of universalization are partial, multiple, and fractured; they are never finished or fully formed."[65] In the end, a teleological model of solidarity that begins with difference and inevitably ends with universalism is not dissimilar from one that presumes universalism from the start. In both cases, universalism is presupposed, and solidarity becomes a magnetic attractor, pulling social actors toward higher principles beyond their blindered worldview. In contrast, a processual model of solidarity relies on social actors to make sense of their own circumstances and deduce universal principles appropriate to the particularities of their experience. As David Featherstone writes, rather than modeling solidarity "as a point of departure," one should instead consider it "as a moment of encounter that creates a horizon for future relationships."[66]

Consistent with the model outlined in the preceding section, this book has shown that unions do not materialize out of the ether; a union willed into existence through immaculate conception may prove inviable. More specifically, the basic tactics that constitute a textbook organizing drive (leafletting, one-on-ones, door-to-door canvassing, informational pickets, confrontations with the boss, etc.) are useful not only because they help the union accomplish its immediate goal of winning a representation election but also because they lay the groundwork for a sustainable organization.[67] Indeed, the very process of organizing can persuade fence-sitters to pursue collective solutions to shared grievances and stiffen supporters' resolve. Effective organizing demonstrates the power of collective action, even as it grows the connective social tissue that enables solidarity. Further, an organizing campaign allows workers to practice risk-taking, mutual aid, group orientation, and a range of other skills and attitudes unfamiliar to most American workers—all of which are critical to a union's future success. Organizing campaigns, therefore, serve a dual function: they permit the group of workers to clear the legal hurdles necessary to achieve representation and to acquire the skills they will need to sustain it. Rick Fantasia argues that a campaign to build a union prefigures the organization it strives to create: "A successful union election represents the institutionalization of that which brought it about."[68] For Fantasia, successful organizing breaks down mistrust and petty resentment between workers, replacing these quotidian divisions with a culture of solidarity. Union elections won through this "praxis of worker solidarity" come to reflect that character once chartered.

Yet, ultimately, in charting the UAW's organizing approach across three election cycles, this book has argued for the centrality of innovation to any solidaristic project. The UAW won in 2024 precisely by breaking the mold that had limited its vision and restricted its ambitions through the previous two elections. As it discovered, solidarity cannot be mass-produced from a preformed mold; building solidarity demands a custom job. In this sense, unionization escapes Marx's often misunderstood distinction between *class-in-itself* and *class-for-itself*: it creates a formal class-based organization, even as it prepares that organization to mobilize around its own class interests. Reversing the classic formula, it is by *acting* as a class that workers affirm their identity as a class-based social category.

Drawing on the four elements of strategic capacity, the UAW has built a type of solidarity with processual, experimental, antiessentialist, reciprocal, agonistic, and generative qualities—transecting (but never transcending) difference. Its durability must still be proved.

Epilogue

B arely a month after its win at VW, the UAW's winning streak was broken by a defeat at Mercedes in Vance, Alabama. This loss undoubtedly
slowed momentum and signals the stiff challenges ahead as the union
seeks to grow its sectoral coverage by organizing transplants, battery plants,
and suppliers. Still, after its groundbreaking win at VW and historic contract with the Big Three, the UAW has not been set back to square one. A
single loss is not enough to cancel the cumulative effects of the Big Three
contracts; the win at VW; smaller victories at Ultium, Blue Oval, and Daimler;
and lower-profile wins at several foreign-owned parts suppliers.

The loss of an election is disappointing, but it reflects the risks inherent
in the UAW's bold momentum-based organizing model. The old UAW was
risk averse, spreading its organizing drives across years and deploying its
organizers methodically and cautiously. In contrast, the new UAW is characterized by risk-taking and creativity, which are both tremendous assets
and sources of some unpredictability. By holding multiple elections in rapid
succession, close on the heels of its historic Big Three victory, the UAW
sought to channel momentum before it dissipated, while ensuring that the
sting of any individual loss was less severe. In years past, holding multiple
elections at disparate sites in the same calendar year would have overwhelmed
the organization's resources. Its willingness to stack the two elections in such
close succession revealed, once again, the degree to which the UAW internal
culture had changed.

Just as at Volkswagen, Mercedes workers were substantially self-organized and fed off momentum from the stand-up strike. Infrastructure on the ground already existed in the form of an independent nonmajority union but was newly invigorated after the UAW strike and VW victory.[1] The UAW has targeted Mercedes for decades, but its prospects seemed remote until the transformative events of 2023–2024 shifted the balance of power suddenly and dramatically. What was once thought impossible now seemed realistic.

However, at Mercedes, the anti-union overreach by management was far more extensive than at VW in 2024, more closely resembling the obstacles the UAW had encountered in 2014. The NLRB found that "Mercedes disciplined employees for discussing unionization at work, prohibited the distribution of union materials and paraphernalia, surveilled employees, discharged union supporters, and forced employees to attend captive audience meetings."[2] Against these obstacles, the UAW faced a stiffer challenge.

The Mercedes campaign was different from VW in other important ways. Alabama is a poorer state than Tennessee, and its economy is more dependent on the automotive industry than Tennessee's. Alabama does not have the tourist draw of a Nashville or Great Smoky Mountains National Park, and the auto industry is arguably more central to the state's self-image. Instead, it is home to Honda, Hyundai, Toyota, and many parts suppliers. At the time of the plant's opening, Alabama spent more on tax breaks and subsidies per job than any other state had spent to court a foreign automaker, approximately $200,000 per job.[3] The EV battery plant is included in the bargaining unit, but the supplier network is not. Alabama paid the salaries of autoworkers at Mercedes while they trained in the 1990s—a gambit to lure investment from auto manufacturers that almost bankrupted the state. Finally, Mercedes enlisted pastors, celebrities, the enormously popular football coach Nick Saban, and other local figures to speak out against the UAW, giving the impression of a more unified opposition to the UAW.

On the positive side, though transnational organizing had generated limited returns in the past, changes to German laws had given the UAW a new opportunity to mobilize legal power against the company, which seemed too tempting to ignore. At Mercedes, the UAW had another weapon in its quiver that was unavailable at VW. Passed in 2021, the German Act on Corporate Due Diligence Obligations in Supply Chains forbids noncompliance with labor protections throughout the supply chain.[4] Still, legislation regulating supply chains faces many of the same challenges as GFAs, and the company roundly ignored these objections.

Any serious analysis of the loss at Mercedes must acknowledge that the UAW has framed its struggle as part of an industry-wide crusade to organize autoworkers rather than a series of isolated campaigns targeting individual employers. That shifts the focus from the outcome of any particular election toward the big picture, which still looks good for the UAW. When one zooms out and considers the state of the labor movement combined with the broader macroeconomic factors that brought us here in the first place, the overall trend still points in an upward direction, even if less definitively than before.

In sum, since its win at VW, the UAW has already shown that transplants are not impervious to organizing. While the South remains challenging terrain, its hostility to unions is not insurmountable. Foreign-owned "transplants" in the American South have long seemed impervious to organizing, but victory at VW breached the dam. Even a string of defeats would not change the fact that the UAW has proved capable of contesting elections in both the North and the South and now has a record of success among domestic firms and transplants. With a toehold in the South, even a temporary setback will not erase the gains of the strike and its aftermath or, for that matter, negate the conditions that made the UAW's victory in Chattanooga possible.

—May 2024, Vance, Alabama

NOTES

PREFACE

1. Ford, General Motors, and Stellantis (formerly Chrysler).

2. Jane McAlevey, *No Shortcuts: Organizing for Power in the New Gilded Age*, illustrated ed. (New York: Oxford University Press, 2016).

3. Abraham Walker, "Unionization at Volkswagen in Chattanooga: A Postmortem," *Labor Studies Journal* 48, no. 2 (2023): 121–48, available at https://doi.org/10.1177/01604 49X231162593.

4. Abe Walker, "Rank-and-File Revolt: Insurgency, Power, and Democracy in the UAW, 2019–25," *Labor History*, Routledge, September 14, 2025, available at https://www .tandfonline.com/doi/abs/10.1080/0023656X.2025.2557988.

5. Abraham Walker, "Third Time's the Charm: Assessing the UAW's Decade-Long Struggle for a Union at Volkswagen," *New Labor Forum* 34, no. 1 (2025): 23–31, available at https://doi.org/10.1177/10957960241299337.

INTRODUCTION

1. Matt Patterson, "Southern Workers Continue to Reject Unions," *Chattanooga Times Free Press* (*CTFP*), January 18, 2014.

2. Mike Pare, "UAW's Effort in High Gear," *CTFP*, September 2, 2013.

3. Marshall Ganz, "Resources and Resourcefulness: Strategic Capacity in the Unionization of California Agriculture, 1959–1966," *American Journal of Sociology* 105, no. 4 (January 2000): 1014–18, available at https://doi.org/10.1086/210398.

4. Marshall Ganz, *Why David Sometimes Wins: Leadership, Organization, and Strategy in the California Farm Worker Movement*, reprint ed. (New York: Oxford University Press, 2010), 252.

5. Marshall Ganz, *Why David Sometimes Wins*.

6. See, for example, Margaret L. Sheng and Iting Chien, "Rethinking Organizational Learning Orientation on Radical and Incremental Innovation in High-Tech Firms," *Journal of Business Research* 69, no. 6 (2016): 2302–8, available at https://doi.org/10.1016/j.jbusres.2015.12.046.

7. Terrence E. Deal and Allan A. Kennedy, *The New Corporate Cultures: Revitalizing the Workplace after Downsizing, Mergers, and Reengineering* (New York: Basic Books, 2000).

8. Ganz, "Resources and Resourcefulness."

9. Within the field of leadership studies, this has become known as a relational model. See Nicholas Clarke, *Relational Leadership: Theory, Practice and Development* (London: Routledge, 2018), available at https://doi.org/10.4324/9781315620435.

10. John Lofland, *Social Movement Organizations: Guide to Research on Insurgent Realities* (New York: Routledge, 1996).

11. Lofland, *Social Movement Organizations*, 171.

12. The "Big Three" are the largest United States–headquartered automobile manufacturers by revenue: General Motors, Ford, and Chrysler (which, since 2021, has operated as a subsidiary of the Dutch multinational Stellantis).

13. Mark Engler and Paul Engler, *This Is an Uprising: How Nonviolent Revolt Is Shaping the Twenty-First Century* (New York: Bold Type Books, 2016).

14. Chris Rhomberg and Steven Lopez, "Understanding Strikes in the 21st Century: Perspectives from the United States," in *Power and Protest: How Marginalized Groups Oppose the State and Other Institutions*, ed. Lisa Leitz, Research in Social Movements, Conflicts and Change 44 (Bingley: Emerald, 2021).

15. Anne Metten, "Rethinking Trade Union Density: A New Index for Measuring Union Strength," *Industrial Relations Journal* 52, no. 6 (2021): 528–49.

16. Zachary Joseph McKenney, "The State of the Union? Transnational Manufacturing and the U.S. Labor Movement" (unpublished diss., December 2018); Stephen J. Silvia, *The UAW's Southern Gamble: Organizing Workers at Foreign-Owned Vehicle Plants*, reprint ed. (Ithaca, NY: ILR Press, 2023); Timothy J. Minchin, *America's Other Automakers: A History of the Foreign-Owned Automotive Sector in the United States* (Athens: University of Georgia Press, 2021).

17. Mary S. Fewtrell et al., "How Much Loss to Follow-Up Is Acceptable in Long-Term Randomised Trials and Prospective Studies?" *Archives of Disease in Childhood* 93, no. 6 (June 1, 2008): 458–61, available at https://doi.org/10.1136/adc.2007.127316.

18. In 2014, 53 percent of workers rejected the union (626 "Yes" to 712 "No" with 89 percent participation). In 2019, 51.8 percent voted against the UAW (833 "Yes" to 776 "No" with 93 percent participation). In 2024, 73 percent voted for the UAW (2,628 "Yes" to 985 "No" with 83.5 percent participation).

19. This chapter title is indebted to Rochelle DuFord, *Solidarity in Conflict: A Democratic Theory* (Stanford, CA: Stanford University Press, 2022).

CHAPTER 1

1. Jane Slaughter, "UAW Says It Will Go 'All In' to Organize Foreign-Owned Auto Plants," *Labor Notes*, January 24, 2011, available at https://www.labornotes.org/2011/01/uaw-says-it-will-go-%E2%80%98all-in%E2%80%99-organize-foreign-owned-auto-plants.

2. The very idea of a postwar contract is, however, contested in the literature. As Nelson Lichtenstein writes, the postwar settlement was a "suspect construct," given the reality that job security was but an aspiration for most workers, and layoffs were frequent, even

during the period of supposed "labor peace." Nostalgia for the 1950s and 1960s only became possible decades later, as the American auto sector confronted an ever more dire existential crisis. In this sense, the notion of a postwar contract "allowed for the creation of a semi-imaginary historical benchmark against which the very real contemporary assaults on unions and key industrial sectors could be measured." Nelson Lichtenstein, *Labor's War at Home: The CIO in World War II* (Cambridge: Cambridge University Press, 1982).

3. Dan Georgakas, Marvin Surkin, and Manning Marable, *Detroit: I Do Mind Dying; A Study in Urban Revolution*, rev. ed. (Cambridge, MA: South End, 1999).

4. Nelson Lichtenstein, *Walter Reuther: The Most Dangerous Man in Detroit* (Urbana: University of Illinois Press, 1997); Thaddeus Russell, *Out of the Jungle: Jimmy Hoffa and the Remaking of the American Working Class* (Philadelphia, PA: Temple University Press, 2003).

5. Beverly Silver, *Forces of Labor: Workers' Movements and Globalization since 1870*, illustrated ed. (Cambridge: Cambridge University Press, 2003).

6. Peter Dicken, *Global Shift: Mapping the Changing Contours of the World Economy* (New York: Guilford, 2015); George Ritzer, *The Globalization of Nothing* (Thousand Oaks, CA: Sage, 2007); Joseph E. Stiglitz, *Globalization and Its Discontents* (New York: W. W. Norton, 2003).

7. Bureau of Labor Statistics, "Annual Employment (Thousands) for NAICS 3361, Motor Vehicle Manufacturing, U.S. Total," 2025, available at https://beta.bls.gov/data Viewer/view/timeseries/IPUEN3361__W200000000;jsessionid=30F7BD24DF2DB5285 66D49D29D1E6085; Bureau of Transportation Statistics, "National Transportation Statistics (Series)," 2019, available at https://doi.org/10.21949/1503663.

8. Timothy J. Minchin, "Look at Detroit: The United Auto Workers and the Battle to Organize Volkswagen in Chattanooga," *Labor History* 60, no. 5 (September 3, 2019): 482–502, available at https://doi.org/10.1080/0023656X.2019.1567700.

9. Barry Eidlin and Jane Slaughter, "After Decades of Corrupt, Antidemocratic Rule, UAW Members Are Finally Electing Their Leaders," *Jacobin*, October 20, 2022, available at https://jacobin.com/2022/10/uaw-reform-leadership-elections-uawd.

10. James Rubenstein, *The Changing U.S. Auto Industry: A Geographical Analysis* (New York: Routledge, 2002).

11. Saul A. Rubinstein and Thomas A. Kochan, *Learning from Saturn: Possibilities for Corporate Governance and Employee Relations* (Ithaca, NY: Cornell University Press, 2001).

12. S&P Global, "EV Manufacturing Race: Which US States Are Taking an Early Lead?" IHS Markit, May 11, 2023, available at https://www.spglobal.com/marketintelligence/en /mi/research-analysis/ev-manufacturing-race-which-us-states-are-taking-early-lead.html.

13. Section 14(b) of the Taft Hartley amendment to the NLRA allows states to pass laws banning "union shop" arrangements, which require workers to join a union as a condition of employment. Such "right-to-work" laws significantly impede organizing efforts by raising barriers to entry. At present, twenty-seven states, including the entire Southeast region, have passed right-to-work laws. Scott Neuman, "Missouri Blocks Right-to-Work Law," NPR.org, August 8, 2018, available at https://www.npr.org/2018/08/08/636568530/mis souri-blocks-right-to-work-law.

14. Alfred Eckes et al., *Globalization and the American South*, ed. James C. Cobb and William Stueck (Athens: University of Georgia Press, 2005).

15. Harold Sirkin, Michael Zinser, and Doug Hohner, *Made in America, Again: Why Manufacturing Will Return to the U.S.* (Boston: Boston Consulting Group, 2011), available at https://www.nist.gov/system/files/documents/2017/05/09/file84471.pdf.

16. Thomas Klier and James M. Rubenstein, "The Evolving Geography of the US Motor Vehicle Industry," in *Handbook of Industry Studies and Economic Geography*, ed.

Frank Giarratani, Geoffrey J. D. Hewings, and Philip McCann (Cheltenham: Edward Elgar, 2013).

17. Michael Goldfield, *The Southern Key: Class, Race, and Radicalism in the 1930s and 1940s* (New York: Oxford University Press, 2020); Ken Riley, "Fighting Racism and Union-Busting in South Carolina," *WorkingUSA* 5, no. 2 (September 1, 2001): 119–30, available at https://doi.org/10.1111/j.1743-4580.2001.00119.x.

18. Strategic missteps by union leaders have done much to maintain this state of affairs. Rather than drawing on their existing structural strength in manufacturing or mobilizing their associational power, union efforts to organize the South were haphazard, insufficiently funded, and disengaged from social activism.

19. Goldfield, *Southern Key*.

20. Timothy J. Minchin, "A World Class Industry: Honda in Alabama and the Rise of the Foreign-Owned Auto Sector," *Journal of Contemporary History* 56, no. 3 (2020): 12.

21. David Anderson and Andrew C. McKevitt, "From 'the Chosen' to the Precariat Southern Workers in Foreign-Owned Factories since the 1980s," in *Reconsidering Southern Labor History: Race, Class, and Power*, ed. Keri Leigh Merritt (Gainesville: University Press of Florida, 2018).

22. Barry Hirsch, David MacPherson, and Wayne Vroman, "Unionstats.Org," 2021, available at http://unionstats.com/MonthlyLaborReviewArticle.htm.

23. Nevada is unique among right-to-work states in that its rate of unionization rivals that of East Coast union strongholds, but this is due largely to its outsize economic dependence on a single, heavily unionized industry: gambling (and the related hospitality trade).

24. Chattanooga Regional Manufacturers Association, *Chattanooga Makes Sense for Manufacturing*, 2019. In their zeal to advertise Chattanooga's pro-business economic climate, city officials pointedly ignored the fact that until recently Chattanooga had been no poster child for unencumbered capitalism. The effort to present it as an anti-union town can only result from collective amnesia. As explored in more detail in Chapter 6, Chattanooga once had a significant union presence. Militant unions once had even bombed bridges and construction sites. See James B. Jones, "Class Consciousness and Worker Solidarity in Urban Tennessee: The Chattanooga Carmen's Strikes of 1899–1917," *Tennessee Historical Quarterly* 52, no. 2 (Summer 1993): 98–112.

25. Michael Porter, *Competitive Strategy: Techniques for Analyzing Industries and Competitors*, illustrated ed. (New York: Free Press, 1998). Michael Porter distinguished between "high-road" firms, which offer well-paid jobs that are secure and satisfying, and "low-road" firms, whose jobs are insecure and unsatisfying, arguing that companies can pick their strategy. High-road firms pay above the legal minimum, offer health insurance and other benefits, and undertake considerable efforts to build motivation and commitment in their employees. Low-road firms are characterized by an exclusive focus on the bottom line, reliance on temporary and contingent employment, and compensation packages that rarely exceed market dictates. Once chosen, a control strategy has important consequences for the kinds of people who are hired, the ways they are trained, and the systems of rewards they face. Cf. Lowell Turner and Michael Fichter, "Perils of the High and Low Roads: Employment Relations in the United States and Germany," in *Labor, Business, and Change in Germany and the United States*, ed. Kirsten S. Wever (Kalamazoo, MI: W. E. Upjohn Institute for Employment Research, 2001), 123–55.

26. Specifically, pay started at $14.50 per hour (compared to $28 for Big Three autoworkers and $42 for VW workers in Germany). One estimate holds that, including benefits, VW's Tennessee operations cost the firm $27 per hour, while in Germany the complete compensation package totaled $67 per hour. A complex bonus system based on individual

and plant-wide metrics incentivized attendance, safety, quality, and productivity at up to 12.5 percent of the base rate, but these payouts were highly contingent and unreliable.

27. Mike Pare and Andy Sher, "State Offered VW $300 Million as UAW Tried to Organize Plant," *CTFP*, April 2, 2014.

28. Good Jobs First, "Subsidy Tracker Search Results," 2020, accessed August 21, 2024, available at https://subsidytracker.goodjobsfirst.org/.

29. Dave Flessner and Mike Pare, "Incentives Boosted to Lure VW Expansion," *CTFP*, July 15, 2014.

30. Mike Pare, "Corker Eyes Idea for Works Council," *CTFP*, July 8, 2014.

31. Lydia DePillis, "Why Volkswagen Is Helping a Union Organize Its Own Plant," *Washington Post*, February 10, 2014, available at https://www.washingtonpost.com/news/wonk/wp/2014/02/10/why-volkswagen-is-helping-a-union-organize-its-own-plant/.

32. Jack Ewing, *Faster, Higher, Farther: How One of the World's Largest Automakers Committed a Massive and Stunning Fraud* (New York: W. W. Norton, 2018).

33. Outside Online, "The 16 Best Places to Live in the U.S.: 2015," August 18, 2015, available at https://www.outsideonline.com/2006426/americas-best-towns-2015.

34. Georg Simmel, *Georg Simmel on Individuality and Social Forms*, ed. Donald N. Levine (Chicago: University of Chicago Press, 1972).

35. John Paul MacDuffie, "Modularity-as-Property, Modularization-as-Process, and 'Modularity'-as-Frame: Lessons from Product Architecture Initiatives in the Global Automotive Industry," *Global Strategy Journal* 3, no. 1 (2013): 8–40, available at https://doi.org/10.1111/j.2042-5805.2012.01048.x.

36. Mike Pare, "Volkswagen Chattanooga Supplier Park Marks 10 Years, Readies for Electric SUV," *CTFP*, October 7, 2020, available at https://www.timesfreepress.com/news/2020/oct/07/volkswagen-chattanooga-supplier-park-marks-10-year/.

37. Volkswagen, "Newsroom: Volkswagen Chattanooga Supplier Park Marks 10 Years of Success," October 6, 2020, available at https://media.vw.com/en-us/releases/1405.

38. Kim Hill and Gregory Burkart, "The Case for Supplier Parks in the Automotive Supply Chain," Area Development, August 12, 2015, available at https://www.areadevelopment.com/Automotive/q3-2015-auto-aero-site-guide/Case-Supplier-Parks-Automotive-Supply-Chain-788899.shtml.

39. Center for Economic Research in Tennessee (CERT), *Tennessee's Automotive Cluster*, 2019, available at https://tnecd.com/wp-content/uploads/2019/05/Automotive-Research-Paper_01-2019-1.pdf.

40. Brian E. Adams, "Migration and Agglomeration among Motor Vehicle Parts Suppliers," December 31, 2015, available at https://www.brianmadams.com/research/mvmigration.pdf.

41. CERT, "Tennessee's Automotive Cluster."

42. William F. Fox and William Hamblen, *Economic Impact of the Volkswagen Assembly Plant in 2012* (University of Tennessee Center for Business and Economic Research, Knoxville, TN, May 31, 2013).

43. Dave Flessner, "These Are the 10 Largest Employers in Chattanooga," *CTFP*, May 1, 2019, available at https://www.timesfreepress.com/news/edge/story/2019/may/01/leaderboard-chattanoogas-10-largest-employers/493092/.

44. Christopher Ludwig, "Supply Chain Conference: Mexican and Intermodal Shockwaves," Automotive Logistics, May 2016, available at https://www.automotivelogistics.media/supply-chain-conference-mexican-and-intermodal-shockwaves/15586.article.

45. Andrew Cox, Joe Sanderson, and Glyn Watson, "Supply Chains and Power Regimes: Toward an Analytic Framework for Managing Extended Networks of Buyer

and Supplier Relationships," *Journal of Supply Chain Management* 37, no. 2 (March 22, 2001): 28–28.

46. Eidlin and Slaughter, "After Decades of Corrupt, Antidemocratic Rule."

47. Ben Hamper, *Rivethead: Tales from the Assembly Line* (New York: Warner Books, 1992).

48. Jane Slaughter, "Concessions Trend Begins at Chrysler," *Labor Notes*, November 20, 1979, available at https://www.labornotes.org/2009/10/labor-history-concessions-trend-begins-chrysler.

49. Chris Kutalik, "Auto Makers Push VEBA Solution for Industry Crisis," *Labor Notes*, August 25, 2007, available at https://labornotes.org/2007/08/auto-makers-push-veba-solution-industry-crisis; Chris Kutalik, "Opposition Swells against Concessionary Auto Contracts," *Labor Notes*, October 29, 2007, available at https://labornotes.org/2007/10/opposition-swells-against-concessionary-auto-contracts.

50. Steven Rattner, *Overhaul: An Insider's Account of the Obama Administration's Emergency Rescue of the Auto Industry* (Boston: Houghton Mifflin Harcourt Trade, 2010); Jane Slaughter, "Unequal Pay for Equal Work," *Labor Notes*, May 17, 2011, available at https://labornotes.org/2011/05/unequal-pay-equal-work; Bill Vlasic, *Once upon a Car: The Fall and Resurrection of America's Big Three Automakers—GM, Ford, and Chrysler*, reprint ed. (New York: William Morrow Paperbacks, 2012). Some commentators dissented from this view, describing two-tier as a reasonable measure to keep the auto companies afloat.

51. Chris Brooks, "On Eve of Union Vote, Chattanooga VW Workers Describe Rampant Workplace Injuries," *Labor Notes*, June 11, 2019, available at https://labornotes.org/2019/06/eve-union-vote-chattanooga-vw-workers-describe-rampant-workplace-injuries.

52. Brooks, "On Eve of Union Vote."

CHAPTER 2

1. The primary reporter assigned to the VW beat was Mike Pare, who covers business for the *Chattanooga Times Free Press*. Like nearly all twenty-first-century newspapers, the *Times Free Press* does not have a dedicated labor reporter on staff.

2. Robert Perrucci, *Japanese Auto Transplants in the Heartland: Corporatism and Community* (New York: Aldine Transaction, 1994), 107.

3. Jeongsuk Joo, "The Impact of the Automobile and Its Culture in the U.S.," *International Area Review* 10, no. 1 (2007): 39–54, https://doi.org/10.1177/223386590701000103.

4. Perrucci, *Japanese Auto Transplants in the Heartland*.

5. Ford's Hapeville, Georgia, plant shut down in 2006, while GM's Doraville plant closed in 2008. As with all of the Big Three's U.S. operations, these plants were unionized, and their closure was consistent with a broader shift to overseas assembly. See Tammy Joyner, "GM Plant Closing: An Era Rolls Away in Doraville," *Atlanta Journal-Constitution*, September 26, 2008; Lindsay Chappell, "VW's Plan for Chattanooga: A Fast and Flexible Launch," *Automotive News: Detroit* 83, no. 6362 (June 1, 2009): 14P; Matt Wilson, "Autoworkers Face 2nd Shift—Displaced Georgians Moved after Earlier Shutdown, Now Part of Idling GM Plant," *CTFP*, June 2, 2009.

6. Mike Pare, "65,000 Seek Work at VW," *CTFP*, November 17, 2009.

7. Felix Bauer, "VW's Fischer: Chattanooga Is My Dream Plant: Frank Fischer," *Automotive News: Detroit* 85, no. 6467 (June 6, 2011): 16.

8. Nelson Lichtenstein, *Labor's War at Home: The CIO in World War II* (Cambridge: Cambridge University Press, 1982).

9. Dave Flessner, "Unions' Numbers, Sway Shrink across Region," *CTFP*, September 6, 2010.

10. Dave Flessner, "VW Plant May Push Up Local Wages," *CTFP*, August 4, 2008.

11. In 2018, the average annual pay for VW production team members was $54,690, compared to the average annual earnings of $45,656 for the Chattanooga area. The relevant Bureau of Labor Statistics category, which includes motor vehicle manufacturing (Cutting, Punching, and Press Machine Setters, Operators, and Tenders, Metal and Plastic), pays $37,310 on average in the Chattanooga metropolitan area. See Bureau of Labor Statistics, "Chattanooga, TN-GA—May 2019 OES Metropolitan and Nonmetropolitan Area Occupational Employment and Wage Estimates," May 2019, available at https://www.bls.gov/oes/current/oes_16860.htm; Bureau of Labor Statistics, "Cutting, Punching, and Press Machine Setters, Operators, and Tenders, Metal and Plastic," May 2019, available at https://www.bls.gov/oes/current/oes514031.htm.

12. David Anderson and Andrew C. McKevitt, "From 'the Chosen' to the Precariat Southern Workers in Foreign-Owned Factories since the 1980s," in *Reconsidering Southern Labor History: Race, Class, and Power*, ed. Keri Leigh Merritt (Gainesville: University Press of Florida, 2018).

13. Chris Brooks, "Volkswagen Reverses Its Union Neutrality, Workers Say," *Labor Notes*, October 28, 2015, available at http://labornotes.org/2015/10/volkswagen-reverses-its-union-neutrality-workers-say.

14. Under the card-check model, unions are tasked with collecting union authorization cards to demonstrate support for unionization. Once a supermajority of the workplace has signed cards authorizing the union as the exclusive employee representative, the company agrees to voluntarily recognize the union and move toward formalizing a contract, eschewing the need for a contested election.

15. This win, discussed in the Introduction, also benefited significantly from the presence of a UAW representative on the board of DaimlerChrysler, of which Freightliner was then a subsidiary. See Michael Fichter, "Building Union Power across Borders: The Transnational Partnership Initiative of IG Metall and the UAW," *Global Labour Journal* 9, no. 2 (May 31, 2018), available at https://doi.org/10.15173/glj.v9i2.3343.

16. Dorothee Benz, "Scaling the Wall of Employer Resistance: The Case for Card Check Campaigns," *New Labor Forum*, no. 3 (1998): 118–28; Adrienne E. Eaton and Jill Kriesky, "Union Organizing under Neutrality and Card Check Agreements," *ILR Review* 55, no. 1 (October 1, 2001): 42–59, available at https://doi.org/10.1177/001979390105500103.

17. Gabe Nelson, "8 VW Workers at Tenn. Plant Allege Misleading UAW Tactics in Organizing Push," *Automotive News*, September 25, 2013, available at https://www.autonews.com/article/20130925/OEM01/130929937/8-vw-workers-at-tenn-plant-allege-misleading-uaw-tactics-in-organizing-push.

18. Mike Pare, "VW Opens Door to Unionization," *CTFP*, March 20, 2013.

19. Fichter, "Building Union Power across Borders," 188.

20. Mike Pare, "VW Deal with UAW Could Take Months," *CTFP*, September 19, 2013.

21. Gabe Nelson, "4 Key VW Decisions Shaped Vote's Course," *Automotive News: Detroit* 88, no. 6609 (February 24, 2014): 1.

22. Mike Pare and Andy Sher, "VW, UAW Pushing Chattanooga Plant Talks," *CTFP*, September 4, 2013.

23. Mike Pare, "VW Official: Chattanooga Works Council Should Come by 'Formal Vote,'" *CTFP*, September 5, 2013.

24. Andrew C. McKevitt, *Consuming Japan: Popular Culture and the Globalizing of 1980s America* (Chapel Hill: University of North Carolina Press, 2017), 111.

25. Stephen Silvia, *Organizing German Automobile Plants in the USA*, Study Der Hans-Böckler-Stiftung 349 (Düsseldorf: Hans-Böckler-Stiftung, 2016), 16.

26. Mike Elk, "Exclusive: Volkswagen Isn't Fighting Unionization—but Leaked Docs Show Right-Wing Groups Are," *In These Times*, November 13, 2013, available at http://inthesetimes.com/working/entry/15876/anti_union_forces_mobilizing_at_chattanoogas_volkswagen_plant.

27. Silvia, *Organizing German Automobile Plants in the USA*, 25. Linking the UAW to industrial collapse is a tried-and-true strategy reemployed by Nissan in Canton, Mississippi, where it described UAW's legacy of "plant closings and layoffs." A local columnist there suggested the UAW aimed to "infiltrate Detroit South and suck it dry." Timothy J. Minchin, "Labor Rights Are Civil Rights: Inter-racial Unionism and the Struggle to Unionize Nissan in Canton, Mississippi," *Labor History* 59, no. 6 (November 2, 2018): 732, available at https://doi.org/10.1080/0023656X.2018.1470213.

28. Mike Pare, "Outside Groups Try to Influence VW Union Choice," *CTFP*, December 13, 2013.

29. Andy Szal, "Mitsubishi Plant Closure Ends Foreign Automakers' UAW Ties," Manufacturing.net, July 28, 2015, available at http://www.manufacturing.net/news/2015/07/mitsubishi-plant-closure-ends-foreign-automakers-uaw-ties.

30. Dan Chapman, "UAW Push to Unionize a Southern Auto Plant Reaches Climax Today with Vote," *Dallas News*, February 14, 2014, available at http://www.dallasnews.com/business/business-headlines/20140214-uaw-push-to-unionize-a-southern-auto-plant-reaches-climax-today-with-vote.ece.

31. Chapman, "UAW Push to Unionize a Southern Auto Plant."

32. Lee Fang, "Anti-UAW Consultant: Defeat 'Invading Union Force' at VW Plant Like Our Confederate Ancestors," *The Nation*, November 14, 2013, available at http://www.thenation.com/article/anti-uaw-consultant-defeat-invading-union-force-vw-plant-our-confederate-ancestors/.

33. Matt Patterson and Julia Tavlas, "UAW Ate Detroit—Chattanooga Could Be Its Next Meal," *CTFP*, June 28, 2013.

34. Robin Smith, "Right to Work—VW May Become Infested by UAW," *CTFP*, February 10, 2014.

35. Andy Sher and Mike Pare, "VW Official Meets Political Leaders, Workers," *CTFP*, November 15, 2013.

36. Phil Williams, "Exclusive: Leaked Emails Raise Questions about Governor's $300M Offer," News Channel 5, April 24, 2014, available at https://www.newschannel5.com/news/newschannel-5-investigates/tennessees-secret-deals/exclusive-leaked-emails-raise-questions-about-governors-300m-offer.

37. Mike Pare, "Corker Eyes Idea for Works Council," *CTFP*, July 8, 2014.

38. During the 1980s, at a time when union struggles were often insular and isolated, Ray Rogers encouraged unions to expand their efforts beyond the narrowly defined workplace and engage multiple stakeholders in an effort to elevate the profile of their struggles. Labeled *corporate campaign*, this strategy supplemented shop-floor activism with outside pressure from shareholders and media-friendly messaging. Bold initiatives at Hormel and Coca-Cola helped establish Rogers as a prominent labor consultant, and his firm Corporate Campaign continues to shape the labor movement. Hardy Green, *On Strike at Hormel: The Struggle for a Democratic Labor Movement* (Philadelphia: Temple University Press, 1991).

Rogers does not advise anti-union initiatives, but his opponents have at times cribbed ideas from his playbook.

39. Mike Pare, "Anti-UAW Website Up for Volkswagen," *CTFP*, September 13, 2013.

40. Mike Pare, "Anti-UAW Petitions Handed Over to VW," *CTFP*, October 5, 2013.

41. Mike Pare, "UAW, VW Didn't Violate Labor Law," *CTFP*, January 24, 2014.

42. Kate Bronfenbrenner, "Final Report: The Effects of Plant Closing or Threat of Plant Closing on the Right of Workers to Organize," *International Publications*, January 1, 1996.

43. Tom Murphy, "VW Chattanooga Plant Now Profitable, Preps for BEV," Wards-Auto, November 22, 2019, available at https://www.wardsauto.com/volkswagen/vw-chattanooga-plant-now-profitable-preps-for-bev.

44. At Nissan in Mississippi, anxiety over temp worker layoffs also hurt the organizing drive. Minchin, "Labor Rights Are Civil Rights."

45. Minchin, "Labor Rights Are Civil Rights," 732.

46. Alexandra Bradbury, "'Employee Engagement' No Substitute for a Union at VW," *Labor Notes*, December 17, 2014, available at https://labornotes.org/2014/12/employee-engagement-no-substitute-union-vw.

47. Diana T. Kurylko, "VW Exec Tackles Quality Woes: Ambitious U.S. Sales Target Hinges on Improving Key Ratings," *Automotive News: Detroit* 86, no. 6483 (September 26, 2011): 3.

48. Diana T. Kurylko, "Here Comes VW, American-Style: To Boost Brand's Sales, New Team Promises More U.S.-Focused Models," *Automotive News: Detroit* 85, no. 6478 (August 22, 2011): 3.

49. Gabe Nelson and Amy Wilson, "An Ultimatum for VW Chattanooga? No Union May Mean No New Product," *Automotive News: Detroit* 87, no. 6574 (June 24, 2013): 3.

50. Mike Pare, "Haslam: VW Workers Happy without Union," *CTFP*, March 21, 2013.

51. Dave Flessner, "Corker Says SUV Vehicle Will Come without UAW," *CTFP*, February 13, 2014 (emphasis added).

52. Mike Pare, "Supplier Question Cranks Up UAW Debate," *CTFP*, February 13, 2014.

53. Holly Webb, "Bo Watson Says VW May Lose State Help If the UAW Is Voted In at Chattanooga Plant," February 10, 2014, available at https://www.chattanoogan.com/2014/2/10/269310/Bo-Watson-Says-VW-May-Lose-State-Help.aspx.

54. One more last-minute surprise attracted far less notice but may have had a similarly decisive impact. Speaking during a closed-door session with Democratic lawmakers, President Obama made offhand comments that were quickly leaked to the press. Though most media accounts implied his remarks had been more extensive, the excerpt that made national headlines involved Obama describing Corker and Haslam as "more concerned about German shareholders than American workers." If accurate, this was a strange maneuver. Haslam and Corker had never indicated that their actions were motivated by concern over VW's shareholders (many of whom are incidentally Americans). But more importantly, by portraying Haslam and Corker as German lackeys, Obama relied on the same rhetorical trope his opponents employed (with greater success), defining the struggle as one between righteous and deserving local interests and manipulative and ideologically suspect foreign ones. See Staff Report, "Obama Backs Union," *CTFP*, February 15, 2014.

55. Mike Pare, "VW Labor Representative Hits Chattanooga Vote," *CTFP*, February 20, 2014.

56. This was not the first time the prospect of an expansion had been used to influence election results. On the eve of the vote, Nissan's head of HR, Gail Newman, declared, "If our environment were to change in such a way that we were unable to remain competi-

tive, then, of course, we would have to reassess investment plans." Timothy J. Minchin, "Showdown at Nissan: The 1989 Campaign to Organize Nissan in Smyrna, Tennessee, and the Rise of the Transplant Sector," *Labor History* 58, no. 3 (May 27, 2017): 396–422, available at https://doi.org/10.1080/0023656X.2017.1262080.

57. In firms with over one thousand employees, the supervisory boards consist of five employer and five worker representatives, with one "neutral" member nominated by employees and approved by shareholders. Shareholders elect half of the board's members, and the works councils select the remaining ones. Two of the five employee representatives must be workers at the firm, and the main German union federation selects two additional employee representatives.

58. Although the German system has stood the test of time and proved resilient, it is not uncontroversial. Some German trade unions view works councils as a check on their power and an incursion into their autonomy. Ulrich Jurgens, *New Worlds of Work: Varieties of Work in Car Factories in the BRIC Countries* (Oxford: Oxford University Press, 2016), 274.

59. In practice, employee-side works councilors are more likely to sign off on management decisions than fight them, and some critics label them little more than "management stooges" empowered to undermine class opposition. Michael Whittall and Rainer Trinczek, "Plant-Level Employee Representation in Germany: Is the German Works Council a Management Stooge or the Representative Voice of the Workforce?" *International Comparative Employee Relations*, November 12, 2019, 125, available at https://www.elgaronline.com/view/edcoll/9781788973212/9781788973212.00017.xml.

60. Again, this is in stark contrast to the U.S. model, where the framers of the NLRA envisioned the state as a neutral arbiter of difference, though eighty years of mission creep have seen the state drifting steadily toward favoring employers. John Rogers Commons, John Bertram Andrews, *Principles of Labor Legislation* (New York: Harper and Bros, 1936); Nelson Lichtenstein, *State of the Union: A Century of American Labor* (Princeton, NJ: Princeton University Press, 2003). There has been renewed interest in the German tripartite model among those who advocate a move toward state-facilitated "social bargaining." Kate Andrias, "The New Labor Law," *Yale Law Journal* 126, no. 1 (2016): 2–100; Kate Andrias, "Social Bargaining in States and Cities: Toward a More Egalitarian and Democratic Workplace Law," *Harvard Law and Policy Review* 12, no. 1 (2017), 19.

61. Stephen J. Silvia, *Holding the Shop Together: German Industrial Relations in the Postwar Era* (Ithaca, NY: Cornell University Press, 2013), 51–61.

62. Though there is no real precedent for works councils in the United States, most legal scholars agree that a works council could only be established at a unionized firm, lest the council violate the prohibition of "company unions."

63. The limits of the works council system were laid bare at the VW plant in Kaluga, Russia—currently, the only other plant in which the works council has not achieved representation. Though works councils exist in Russia, laws in that country grant ultimate authority to trade unions while denying works councils access to information and preventing them from negotiating with employers. Accustomed to works councils with a much broader purview, managers at the Kaluga plant proved incapable of devising a model consistent with these restrictions.

64. In India, for example, VW management tried to modify a works committee (the standard mode of representation in India) as a proxy for the works council. Jurgens, *New Worlds of Work*.

65. Matthew Finkin and Thomas Kochan, "The Volkswagen Way to Better Labor-Management Relations," *CTFP*, January 23, 2014.

66. Stephen J. Silvia, *The UAW's Southern Gamble: Organizing Workers at Foreign-Owned Vehicle Plants*, reprint ed. (Ithaca, NY: ILR Press, 2023).

67. Gary Casteel, "Works Council Crucial to Success, Growth of UAW Here," *CTFP*, July 14, 2013.

68. Ben Klayman, "UAW Seeks VW's Blessing to Represent Tennessee Workers," Reuters, September 13, 2013, available at https://www.reuters.com/article/us-autos-uaw-vw-idUSBRE98C03920130913.

69. Pare, "Haslam: VW Workers Happy without Union."

70. Chris Brooks, "Volkswagen in Tennessee: Productivity's Price," *Labor Notes*, March 12, 2015, available at https://labornotes.org/2015/03/volkswagen-tennessee-productivitys-price.

71. Anonymous informant, interview with author, September 21, 2016.

72. Private Facebook group post.

73. Stephen J. Silvia, "The United Auto Workers' Attempts to Unionize Volkswagen Chattanooga," *ILR Review* 71, no. 3 (2018): 600–624, available at https://doi.org/10.1177/0019793917723620.

74. Anonymous informant, interview with author, August 20, 2017.

75. Anonymous informant, interview with author, September 21, 2016.

76. Anonymous informant, interview with author, September 21, 2016.

77. Dale Hathaway, *Allies across the Border* (Cambridge, MA: South End, 2000); Jamie K. McCallum, *Global Unions, Local Power: The New Spirit of Transnational Labor Organizing* (Ithaca, NY: Cornell University Press, 2013); Peter Waterman and Jane Wills, eds., *Place, Space and the New Labour Internationalisms* (Oxford: Wiley-Blackwell, 2002).

78. Bengt Larsson, "Obstacles to Transnational Trade Union Cooperation in Europe—Results from a European Survey," *Industrial Relations Journal* 43, no. 2 (2012): 152–70, available at https://doi.org/10.1111/j.1468-2338.2012.00666.x.

79. Lance Compa, "When in Rome: The Exercise of Power by Foreign Multinational Companies in the United States," *Transfer: European Review of Labour and Research* 20, no. 2 (May 1, 2014): 273, available at https://doi.org/10.1177/1024258914526105.

80. Meric S. Gertler and Tara Vinodrai, "Learning from America? Knowledge Flows and Industrial Practices of German Firms in North America," *Economic Geography* 81, no. 1 (2005): 49, available at https://doi.org/10.1111/j.1944-8287.2005.tb00254.x.

81. Marc de Smidt and Egbert Wever, *The Corporate Firm in a Changing World Economy: Case Studies in the Geography of Enterprise* (London: Routledge, 2012).

82. Mitchell Smith, interview with author, February 18, 2017.

83. Anonymous informant, interview with author, September 21, 2016.

84. Bauer, "VW's Fischer."

85. Bauer, "VW's Fischer."

86. Chappell, "VW's Plan for Chattanooga."

87. Bauer, "VW's Fischer."

88. Over the last thirty years, most automobile firms have reconfigured themselves as holding companies for distributed global networks, buying up boutique brands while allowing their own subsidiaries a measure of autonomy. VW has also moved toward decentralization but has been somewhat slower to evolve and remained a traditional, hierarchical firm well into the twenty-first century. See Ludger Pries, "Volkswagen in the 1990s: Accelerating from a Multinational to a Transnational Automobile Company," in *Globalization or Regionalization of the European Car Industry?* ed. Michel Freyssenet, Koichi Shimizu, and Giuseppe Volpato (Houndmills: Palgrave Macmillan, 2013), 51–72.

89. Ulrich Jurgens, "Implanting Change: The Role of 'Indigenous Transplants' in Transforming the German Productive Model," in *Between Imitation and Innovation: The Transfer and Hybridization of Productive Models in the International Automobile Industry* ed. Robert Boyer et al. (Oxford: Oxford University Press, 1998), 339.

90. Jurgens, "Implanting Change."

91. Gabe Nelson, "U.S. Execs Lead VW's Quality Quest in N.A.: Fixes Needed? No Need to Wait for OK," *Automotive News: Detroit* 87, no. 6559 (March 11, 2013): 6.

92. Jefferson Cowie, *Capital Moves: RCA's Seventy-Year Quest for Cheap Labor* (Ithaca, NY: Cornell University Press, 2019).

93. Peter A. Hall and David W. Soskice, *Varieties of Capitalism: The Institutional Foundations of Comparative Advantage* (Oxford: Oxford University Press, 2001).

94. Thomas W. Dunfee and Timothy L. Fort, "Corporate Hypergoals, Sustainable Peace, and the Adapted Firm," *Vanderbilt Journal of Transnational Law* 36, no. 2 (2003): 567.

95. Indeed, there are growing signs of a reverse boomerang effect, in which European firms will eventually "bring the war home," importing U.S.-style labor relations to their native-born workplaces, further complicating their cultural identities.

CHAPTER 3

1. Robert R. Alford, *The Craft of Inquiry: Theories, Methods, Evidence* (New York: Oxford University Press, 1998).

2. Timothy J. Minchin, "Showdown at Nissan: The 1989 Campaign to Organize Nissan in Smyrna, Tennessee, and the Rise of the Transplant Sector," *Labor History* 58, no. 3 (May 27, 2017): 396–422, available at https://doi.org/10.1080/0023656X.2017.1262080; Timothy J. Minchin, "Labor Rights Are Civil Rights: Inter-racial Unionism and the Struggle to Unionize Nissan in Canton, Mississippi," *Labor History* 59, no. 6 (November 2, 2018), available at https://doi.org/10.1080/0023656X.2018.1470213; Stephen J. Silvia, *The UAW's Southern Gamble: Organizing Workers at Foreign-Owned Vehicle Plants*, reprint ed. (Ithaca, NY: ILR Press, 2023).

3. Kate Bronfenbrenner and Robert Hickey, "Changing to Organize," in *Rebuilding Labor*, by Ruth Milkman and Kim Voss (Ithaca, NY: Cornell University Press, 2004), 17–61.

4. Kate Bronfenbrenner and Tom Juravich, "It Takes More than House Calls: Organizing to Win with a Comprehensive Union-Building Strategy," in *Organizing to Win: New Research on Union Strategies* ed. Kate Bronfenbrenner et al. (Ithaca, NY: ILR Press, 1998), 19–36.

5. Neil Bucklew et al., "Revisiting U.S. Labor Law as a Restriction to Works Councils: A Key for U.S. Global Competitiveness," *Drake Law Review*, January 1, 2018, available at https://researchrepository.wvu.edu/law_faculty/20.

6. In 2014, the UAW referred to its OC as an "leadership council." Because this name change was purely semantic, use the more conventional "organizing committee."

7. Chris Bohner, "New 2023 Data on Union Membership and Finances," *Radish Research* (blog), Substack, April 25, 2024, available at https://radishresearch.substack.com/p/new-2023-data-on-union-membership.

8. Chris Bohner, "The Labor Movement's 'Business Unionism' Has Transformed into 'Finance Unionism,'" February 5, 2023, available at https://jacobin.com/2023/02/finance-unionism-union-density-decline-american-labor-movement-mass-organizing.

9. Michael Gilliand, Interview with author, December 28, 2018.

10. Mike Pare, "Corker Eyes Idea for Works Council," *CTFP*, July 8, 2014.

11. Volkswagen Group of America and United Auto Workers, "Agreement for a Representation Election," Chattanooga, TN, January 27, 2014 [emphasis added].

12. Gabe Nelson, "4 Key VW Decisions Shaped Vote's Course," *Automotive News: Detroit* 88, no. 6609 (February 24, 2014): 1.

13. Stephen Silvia, *Organizing German Automobile Plants in the USA*, Study Der Hans-Böckler-Stiftung 349 (Düsseldorf: Hans-Böckler-Stiftung, 2016).

14. One notable exception was the organizing drive in Oxford, Mississippi, in which another Nissan plant in Tennessee allowed for a transplant-to-transplant comparison. Minchin, "Labor Rights Are Civil Rights."

15. David Barkholz, "VW Workers Saw Little Gain in UAW Vote," *Automotive News: Detroit* 88, no. 6609 (February 24, 2014): 32.

16. See, for example, Matthew Patterson, "UAW Imports Anti-Worker Thuggery to Tennessee." *Washington Examiner* (February 2, 2014).

17. Richard Barry Freeman and Joel Rogers, *What Workers Want* (Ithaca, NY: Cornell University Press, 2006).

18. Anonymous informant, interview with author, July 17, 2017.

19. Anonymous informant, interview with author, August 20, 2017.

20. Anonymous informant, interview with author, August 20, 2017.

21. Anonymous informant, interview with author, June 6, 2019.

22. Alexandra Bradbury, "'Employee Engagement' No Substitute for a Union at VW," *Labor Notes*, December 17, 2014, available at https://labornotes.org/2014/12/employee-engagement-no-substitute-union-vw.

23. Mike Pare, "VW Unionizing Effort 'Low Key, Low Budget,'" *CTFP*, September 18, 2013.

24. Mike Pare, "Outside Groups Try to Influence VW Union Choice," *CTFP*, December 13, 2013.

25. Chris Brooks, "Organizing Volkswagen: A Critical Assessment," *WorkingUSA* 19 (September 1, 2016), available at https://doi.org/10.1111/wusa.12249.

26. Mike Pare, "Union Criticizes 'Outside Groups'—More Anti-UAW Billboards Going Up in Chattanooga," *CTFP*, February 7, 2014.

27. Pare, "Union Criticizes 'Outside Groups.'"

28. The organizing committee called itself a "leadership council" in 2014. I use the more conventional term to maintain consistency.

29. Anonymous Informant, Interview with author, September 21, 2016.

30. Anonymous Informant, interview with author, January 6, 2017.

31. Anonymous Informant, Interview with author, September 6, 2018.

32. Kate Bronfenbrenner et al., *Organizing to Win: New Research on Union Strategies* (Ithaca, NY: Cornell University Press, 1998).

33. Anonymous Informant, Interview with author, September 21, 2016.

34. Mike Elk, "Nissan Attacked for One of 'Nastiest Anti-union Campaigns' in Modern US History," *The Guardian*, August 1, 2017, sec. US News, available at https://www.theguardian.com/us-news/2017/aug/01/nissan-mississippi-union-vote.

35. Barbara S. Griffith, *The Crisis of American Labor: Operation Dixie and the Defeat of the CIO* (Philadelphia: Temple University Press, 1988), 29.

36. Marc Doussard and Brad R. Fulton, "Organizing Together: Benefits and Drawbacks of Community-Labor Coalitions for Community Organizations," *Social Service Review* 94, no. 1 (March 1, 2020): 36–74, available at https://doi.org/10.1086/707568; Bruce Nissen, "The Effectiveness and Limits of Labor-Community Coalitions: Evidence from South Florida," *Labor Studies Journal* 29, no. 1 (March 1, 2004): 67–88, available at https://doi

.org/10.1177/0160449X0402900105; Lowell Turner and Daniel B. Cornfield, *Labor in the New Urban Battlegrounds: Local Solidarity in a Global Economy* (Ithaca, NY: Cornell University Press, 2007).

37. Marissa Brookes, "Varieties of Power in Transnational Labor Alliances: An Analysis of Workers' Structural, Institutional, and Coalitional Power in the Global Economy," *Labor Studies Journal* 38, no. 3 (September 1, 2013): 181–200, available at https://doi.org/10.1177/0160449X13500147.

38. Amanda Tattersall, *Power in Coalition: Strategies for Strong Unions and Social Change* (Ithaca, NY: ILR Press, 2010).

39. Nissen, "Effectiveness and Limits of Labor-Community Coalitions."

40. Brooks, "Organizing Volkswagen."

41. Michael Gilliand, interview with author, December 17, 2017.

42. See Robin D. G. Kelley, *Hammer and Hoe: Alabama Communists during the Great Depression*, Twenty-fifth anniversary edition (The University of North Carolina Press, 2015), https://search.ebscohost.com/login.aspx?direct=true&scope=site&db=nlebk&db=nlabk&AN=978206.

43. Robert Korstad and Nelson Lichtenstein, "Opportunities Found and Lost: Labor, Radicals, and the Early Civil Rights Movement," *Journal of American History* 75, no. 3 (December 1, 1988): 786–811, available at https://doi.org/10.2307/1901530.

44. Rick Halpern, "Organized Labour, Black Workers and the Twentieth-Century South: The Emerging Revision," *Social History* 19, no. 3 (1994): 379.

45. Henry McKiven, *Iron and Steel: Class, Race, and Community in Birmingham, Alabama, 1875–1920*, new ed. (Chapel Hill: University of North Carolina Press, 1991).

46. Robert H. Zieger, *For Jobs and Freedom: Race and Labor in America since 1865* (Lexington: University Press of Kentucky, 2007), available at https://www.jstor.org/stable/j.ctt5vkkh1.

47. Karsten Hulsemann, "Greenfields in the Heart of Dixie: How the American Auto Industry Discovered the South," in *The Second Wave: Southern Industrialization from the 1940s to the 1970s*, ed. Philip Scranton (Athens: University of Georgia Press, 2001), 228.

48. Mike Elk, "After Ten-Year Battle, a Younger Generation Leads the Way at Volkswagen," *American Prospect*, April 19, 2024, available at https://prospect.org/api/content/a07da780-fdc6-11ee-9def-12163087a831/.

49. U.S. Census Bureau, "U.S. Census Bureau QuickFacts: Chattanooga City, Tennessee; United States," July 1, 2019, available at https://www.census.gov/quickfacts/fact/table/chattanoogacitytennessee,US/PST045218.

50. Anonymous informant, interview with author, August 20, 2017.

51. At the time of the microunit election, 163 skilled machinists were employed, out of 1,400 workers plant-wide. Expansions have since changed these numbers dramatically, but the overall balance remains roughly the same.

52. Anonymous informant, interview with author, August 20, 2017.

53. Mark Mix, "Union Push: VW Workers, Be Wary of Secret Deals," *CTFP*, April 13, 2013.

54. Nick Bunkley, "Bob King's Final Battle: UAW Uses Multifront Push to Organize Imports," *Automotive News: Detroit* 88, no. 6591 (October 21, 2013): 1.

55. UAW, "UAW Statement on VW Works Council in Chattanooga," September 6, 2013, available at http://uaw.org/uaw-statement-on-vw-works-council-in-chattanooga/.

56. Lydia DePillis, "Auto Union Loses Historic Election at Volkswagen Plant in Tennessee," *Washington Post*, February 14, 2014, available at https://www.washingtonpost.com

/news/wonk/wp/2014/02/14/united-auto-workers-lose-historic-election-at-chattanoo
ga-volkswagen-plant/.

57. Anonymous informant, interview with author, July 15, 2017.

58. Anonymous informant, interview with author, July 15, 2017.

59. Private Facebook group post.

60. Private Facebook group post.

61. Anonymous informant, interview with author, July 15, 2017.

62. Anonymous informant, interview with author, July 15, 2017.

63. Private Facebook group post.

64. Anonymous informant, interview with author, August 20, 2017.

65. Ron Martin, "Economic Geography and the New Discourse of Regional Competitiveness," in *Economic Geography: Past, Present and Future* ed. Sharmistha Bagchi-Sen and Helen Lawton-Smith (London: Routledge, 2006).

66. "Fighting the Union in a 'Union Friendly' Company: The AT&T/NCR Case," *Labor Studies Journal* 23, no. 3 (September 1, 1998): 3–32, available at https://doi.org/10.1177/0160449X9802300301.

CHAPTER 4

1. Global framework agreements have been referred to variously as international framework agreements and transnational collective agreements. As these terms are functionally identical, I exclusively use the GFA nomenclature. Substitutions in quoted passages are indicated by square brackets as appropriate.

2. Andreas Bieler and Ingemar Lindberg, eds., *Global Restructuring, Labour and the Challenges for Transnational Solidarity* (New York: Routledge, 2010); Nathan Lillie and Miguel Martínez Lucio, "Rollerball and the Spirit of Capitalism: Competitive Dynamics within the Global Context, the Challenge to Labour Transnationalism, and the Emergence of Ironic Outcomes," *Critical Perspectives on International Business* 8, no. 1 (January 1, 2012): 74–92, available at https://doi.org/10.1108/17422041211197576; Marcel van der Linden, "Transnationalizing American Labor History," *Journal of American History* 86, no. 3 (1999): 1078–92, available at https://doi.org/10.2307/2568606.

3. Ian Greer and Marco Hauptmeier, "Management Whipsawing: The Staging of Labor Competition under Globalization," *ILR Review* 69, no. 1 (January 1, 2016): 31, available at https://doi.org/10.1177/0019793915602254.

4. Dimitris Stevis and Terry Boswell, *Globalization and Labor: Democratizing Global Governance* (Lanham, UK: Rowman and Littlefield, 2008).

5. Kate Bronfenbrenner, *Global Unions: Challenging Transnational Capital through Cross-Border Campaigns* (Ithaca, NY: Cornell University Press, 2007).

6. Mark S. Anner, *Solidarity Transformed: Labor Responses to Globalization and Crisis in Latin America* (Ithaca, NY: Cornell University Press, 2011); Mark Anner et al., "The Industrial Determinants of Transnational Solidarity: Global Interunion Politics in Three Sectors," *European Journal of Industrial Relations* 12, no. 1 (March 2006): 7–27.

7. Jimmy Donaghey et al., "From Employment Relations to Consumption Relations: Balancing Labor Governance in Global Supply Chains," *Human Resource Management* 53, no. 2 (2014): 229–52, available at https://doi.org/10.1002/hrm.21552.

8. Marissa Brookes and Jamie K. McCallum, "The New Global Labour Studies: A Critical Review," *Global Labour Journal* 8, no. 3 (September 30, 2017), available at https://doi.org/10.15173/glj.v8i3.3000.

9. Ian Greer and Marco Hauptmeier, "Political Entrepreneurs and Co-Managers: Labour Transnationalism at Four Multinational Auto Companies," *British Journal of Industrial Relations* 46, no. 1 (2008): 76–97, https://doi.org/10.1111/j.1467-8543.2007.00667.x.

10. Christopher L. Pallas, "Inverting the Boomerang: Examining the Legitimacy of North–South–North Campaigns in Transnational Advocacy," *Global Networks* 17, no. 2 (2017): 281–99, https://doi.org/10.1111/glob.12129.

11. Peter Evans, "National Labor Movements and Transnational Connections: Global Labor's Evolving Architecture under Neoliberalism," *Global Labour Journal* 5, no. 3 (2014), available at https://doi.org/10.15173/glj.v5i3.2283.

12. Rémi Bourguignon, Pierre Garaudel, and Simon Porcher, "Global Framework Agreements and Trade Unions as Monitoring Agents in Transnational Corporations," *Journal of Business Ethics* 165, no. 3 (September 1, 2020): 517–33, available at https://doi.org/10.1007/s10551-019-04115-w.

13. Giovanni Arrighi, "Spatial and Other Fixes of Historical Capitalism," *Journal of World-Systems Research* 10, no. 2 (2004): 527–39; Beverly Silver, *Forces of Labor: Workers' Movements and Globalization since 1870*, illustrated ed. (Cambridge: Cambridge University Press, 2003).

14. Michael Fichter, Markus Helfen, and Jörg Sydow, "Employment Relations in Global Production Networks: Initiating Transfer of Practices via Union Involvement," *Human Relations* 64, no. 4 (April 1, 2011): 599–622, available at https://doi.org/10.1177/0018726710396245.

15. Christian Lévesque et al., "Corporate Social Responsibility and Worker Rights: Institutionalizing Social Dialogue through International Framework Agreements," *Journal of Business Ethics* 153 (November 1, 2018), available at https://doi.org/10.1007/s10551-016-3370-9.

16. Glynne Williams, Steve Davies, and Crispen Chinguno, "Subcontracting and Labour Standards: Reassessing the Potential of International Framework Agreements," *British Journal of Industrial Relations* 53, no. 2 (2015): 181–203, available at https://doi.org/10.1111/bjir.12011.

17. Jane Wills, "Bargaining for the Space to Organize in the Global Economy: A Review of the Accor-IUF Trade Union Rights Agreement," *Review of International Political Economy* 9, no. 4 (2002): 675–700.

18. Jörg Sydow et al., "Implementation of Global Framework Agreements: Towards a Multi-organizational Practice Perspective," *Transfer: European Review of Labour and Research* 20, no. 4 (November 1, 2014): 489–503, available at https://doi.org/10.1177/1024258914546270.

19. Lillie and Martínez Lucio, "Rollerball and the Spirit of Capitalism"; Peter Wad, "'Due Diligence' at APM-Maersk," in *Global Unions: Challenging Transnational Capital through Cross-Border Campaigns*, ed. Kate Bronfenbrenner (Ithaca, NY: Cornell University Press, 2007).

20. Christina Niforou, "International Framework Agreements and Industrial Relations Governance: Global Rhetoric versus Local Realities," *British Journal of Industrial Relations* 50, no. 2 (2012): 367, available at https://doi.org/10.1111/j.1467-8543.2011.00851.x.

21. Niforou, "International Framework Agreements and Industrial Relations Governance," 370.

22. Lillie and Martínez Lucio, "Rollerball and the Spirit of Capitalism," 84.

23. Lévesque et al., "Corporate Social Responsibility and Worker Rights."

24. Markus Helfen and Michael Fichter, "Building Transnational Union Networks across Global Production Networks: Conceptualising a New Arena of Labour–Management Relations," *British Journal of Industrial Relations* 51, no. 3 (2013): 553–76, available at https://doi.org/10.1111/bjir.12016.

25. Dimitris Stevis and Terry Boswell, "International Framework Agreements," in Bronfenbrenner, *Global Unions*, 184.

26. Torsten Müller, Hans-Wolfgang Platzer, and Stefan Rüb, "European Collective Agreements at Company Level and the Relationship between EWCs and Trade Unions—Lessons from the Metal Sector," *Transfer: European Review of Labour and Research* 17, no. 2 (May 1, 2011): 217–28, available at https://doi.org/10.1177/1024258911401448.

27. Stevis and Boswell, "International Framework Agreements."

28. Stevis and Boswell, "International Framework Agreements," 193.

29. Subhabrata Bobby Banerjee, "Transnational Power and Translocal Governance: The Politics of Corporate Responsibility," *Human Relations* 71, no. 6 (June 1, 2018): 796–821, available at https://doi.org/10.1177/0018726717726586; Cedric Dawkins, "An Agonistic Notion of Political CSR: Melding Activism and Deliberation," *Journal of Business Ethics* 170 (April 1, 2021), available at https://doi.org/10.1007/s10551-019-04352-z; Michael Fichter and Jamie K. McCallum, "Implementing Global Framework Agreements: The Limits of Social Partnership," *Global Networks* 15, no. s1 (2015): S65–85, available at https://doi.org/10.1111/glob.12088.

30. Anner et al., "Industrial Determinants of Transnational Solidarity."

31. Christina Niforou, "International Framework Agreements and the Democratic Deficit of Global Labour Governance," *Economic and Industrial Democracy* 35, no. 2 (May 1, 2014): 367–86, available at https://doi.org/10.1177/0143831X13484815; Santanu Sarkar and Sarosh Kuruvilla, "Constructing Transnational Solidarity: The Role of Campaign Governance," *British Journal of Industrial Relations* 58, no. 1 (2020): 27–49, available at https://doi.org/10.1111/bjir.12465.

32. Peter Waterman, *Globalization, Social Movements, and the New Internationalism* (London: Continuum, 2001), 315. Nonetheless, counterexamples abound. More commonly, GFA-based global initiatives struggle to make an impact. For instance, despite considerable resources devoted by the German union ver.di to assist the Communications Workers of America with organizing efforts at T-Mobile, the impact on the ground has been negligible. Marissa Brookes, "Varieties of Power in Transnational Labor Alliances: An Analysis of Workers' Structural, Institutional, and Coalitional Power in the Global Economy," *Labor Studies Journal* 38, no. 3 (September 1, 2013): 190, available at https://doi.org/10.1177/0160449X13500147.

33. Juliane Reinecke and Jimmy Donaghey, "Political CSR at the Coalface—the Roles and Contradictions of Multinational Corporations in Developing Workplace Dialogue," *Journal of Management Studies* 58, no. 2 (2021): 458, available at https://doi.org/10.1111/joms.12585.

34. Andrew Cumbers and Paul Routledge, "The Entangled Geographies of Transnational Labour Solidarity," in *Missing Links in Labour Geography* (London: Routledge, 2010), ed. Ann Cecilie Bergene, Sylvi B. Endersen and Hege Merete Knutsen; Michael Fichter et al., "Globalising Labour Relations: On Track with Framework Agreements?" (SSRN Scholarly Paper, Rochester, NY, July 12, 2013), available at https://doi.org/10.2139/ssrn.2292894; Owen E. Hernstradt, "Are International Framework Agreements a Path to Corporate Social Responsibility?," *University of Pennsylvania Journal of Business and Employment Law* 10 (2007): 187.

35. Helfen and Fichter, "Building Transnational Union Networks across Global Production Networks."

36. Stephen Mustchin and Miguel Martínez Lucio, "Transnational Collective Agreements and the Development of New Spaces for Union Action: The Formal and Informal Uses of International and European Framework Agreements in the UK," *British Journal of Industrial Relations* 55, no. 3 (2017): 599, available at https://doi.org/10.1111/bjir.122 44.

37. Markus Helfen and Jörg Sydow, "Negotiating as Institutional Work: The Case of Labour Standards and International Framework Agreements," *Organization Studies* 34, no. 8 (August 1, 2013): 1073–98, available at https://doi.org/10.1177/0170840613492072.

38. Fichter and McCallum, "Implementing Global Framework Agreements," 66.

39. Richard Hyman, "Trade Unions and the Politics of the European Social Model," *Economic and Industrial Democracy* 26 (February 1, 2005), available at https://doi.org /10.1177/0143831X05049401.

40. Fichter and McCallum, "Implementing Global Framework Agreements," 267.

41. Marc-Antonin Hennebert, Isabelle Roberge-Maltais, and Urwana Coiquaud, "The Effectiveness of International Framework Agreements as a Tool for the Protection of Workers' Rights: A Metasynthesis," *Industrial Relations Journal* 54, no. 3 (2023): 251, available at https://doi.org/10.1111/irj.12398.

42. Mustchin and Lucio, "Transnational Collective Agreements," 599.

43. José Ricardo Ramalho and Marco Aurélio Santana, "VW's Modular System and Workers' Organization in Resende, Brazil," *International Journal of Urban and Regional Research* 26, no. 4 (2002): 756–66, available at https://doi.org/10.1111/1468-2427.00416.

44. Chris Bolsmann, "Contesting Labor Internationalism: The 'Old' Trapped in the 'New' in Volkswagen's South African Plant," *Labor Studies Journal* 35, no. 4 (December 1, 2010): 520–39, available at https://doi.org/10.1177/0160449X10365560; Geoffrey Wood and Pralene Mahabir, "South Africa's Workplace Forum System: A Stillborn Experiment in the Democratisation of Work?" *Industrial Relations Journal* 32, no. 3 (2001): 230–43, available at https://doi.org/10.1111/1468-2338.00195.

45. Marco Hauptmeier and Matt Vidal, *Comparative Political Economy of Work* (New York: Bloomsbury, 2014), 181.

46. Mona Aranea, Sergio González Begega, and Holm-Detlev Köhler, "The European Works Council as a Management Tool to Divide and Conquer: Corporate Whipsawing in the Steel Sector," *Economic and Industrial Democracy* 42, no. 3 (August 1, 2021): 873–91, available at https://doi.org/10.1177/0143831X18816796; Carola M. Frege, "A Critical Assessment of the Theoretical and Empirical Research on German Works Councils," *British Journal of Industrial Relations* 40, no. 2 (2002): 221–48, available at https://doi .org/10.1111/1467-8543.00230; Michael Whittall et al., "Workplace Trade Union Engagement with European Works Councils and Transnational Agreements: The Case of Volkswagen Europe," *European Journal of Industrial Relations* 23, no. 4 (December 1, 2017): 397–414, available at https://doi.org/10.1177/1721727X17699444.

47. Ulrich Jurgens, *New Worlds of Work: Varieties of Work in Car Factories in the BRIC Countries* (Oxford: Oxford University Press, 2016), 287.

48. Stephen Silvia, *Organizing German Automobile Plants in the USA*, Study Der Hans-Böckler-Stiftung 349 (Düsseldorf: Hans-Böckler-Stiftung, 2016).

49. Adelheid Hege and Christian Dufour, "Decentralization and Legitimacy in Employee Representation: A Franco-German Comparison," *European Journal of Industrial Relations* 1, no. 1 (March 1, 1995): 83–99, available at https://doi.org/10.1177/095968019511006.

50. Jefferson Cowie, "National Struggles in a Transnational Economy: A Critical Analysis of US Labor's Campaign against NAFTA," *Labor Studies Journal* 21, no. 4 (1997): 3–32.

51. Anner et al., "Industrial Determinants of Transnational Solidarity."

52. Kevin Young and Diana C. Sierra Becerra, "How 'Partnership' Weakens Solidarity: Colombian GM Workers and the Limits of UAW Internationalism," *WorkingUSA* 17, no. 2 (2014): 239–60, available at https://doi.org/10.1111/wusa.12109.

53. After the 2014 election ended in defeat, the UAW redoubled its efforts to build a cross-border alliance with IGM. To its credit, starting in 2017, the UAW, with cooperation from IGM, embarked on an ambitious transnational labor initiative, headed by independent journalist Carsten Huebner. This initiative was nothing short of groundbreaking, given that American auto plants do not have a long-standing tradition of international solidarity. Even in Europe, where cross-border campaigns have a long history, it is often difficult to persuade workers that their fate is bound up with others. Several workers did have the opportunity to attend an annual works council meeting in Stuttgart as independent observers. This is notable in that cross-border campaigns often lack face-to-face contact between workers. But even these sessions did not persuade the attendees that the works council would improve their fortunes dramatically. Absent a real commitment by foreign unions to engage in job actions, international solidarity drifted inevitably toward the purely symbolic realm.

54. Michael Fichter, *Trade Unions in Transformation: Transnationalizing Unions: The Case of the UAW and the IG Metall* (Berlin: Friedrich Ebert Stiftung, 2017), available at https://library.fes.de/pdf-files/iez/14342.pdf.

55. Brookes, "Varieties of Power in Transnational Labor Alliances," 201.

56. Given the limitations of time and space, this draft is current as of March 2024. A final version will likely incorporate new and emergent developments, which remained unresolved at the time of writing.

57. IG Metall, "Background to International Framework Agreements in the IMF," September 26, 2006, 8, available at https://library.fes.de/pdf-files/gurn/00251.pdf.

58. Michael Fichter, "Building Union Power across Borders: The Transnational Partnership Initiative of IG Metall and the UAW," *Global Labour Journal* 9, no. 2 (May 31, 2018), available at https://doi.org/10.15173/glj.v9i2.3343.

59. Neil Bucklew et al., "Revisiting U.S. Labor Law as a Restriction to Works Councils: A Key for U.S. Global Competitiveness," *Drake Law Review*, January 1, 2018, available at https://researchrepository.wvu.edu/law_faculty/20.

60. Wills, "Bargaining for the Space to Organize in the Global Economy."

61. Nikolaus Hammer and Lene Riisgaard, "Labour and Segmentation in Value Chains," in *Putting Labour in its Place: Labour Process Analysis and Global Value Chains*, ed. Kirsty Newsome et al. (London: Palgrave, 2015), 83–99.

62. Niforou, "International Framework Agreements and Industrial Relations Governance."

63. Konstantinos Papadakis and Dominique Bé, *Cross-Border Social Dialogue and Agreements: An Emerging Global Industrial Relations Framework?* (International Institute for Labour Studies, 2008), available at http://www.ilo.org/public/english/bureau/inst/download/cross.pdf.

64. Markus Helfen, Elke Schubler, and Dimitris Stevis, "Translating European Labor Relations Practices to the United States through Global Framework Agreements? German and Swedish Multinationals Compared," *ILR Review* 69, no. 3 (2016): 631–55.

65. Niforou, "International Framework Agreements and Industrial Relations Governance."

66. Hennebert, Roberge-Maltais, and Coiquaud, "Effectiveness of International Framework Agreements"; Miguel Martínez Lucio, "Dimensions of Internationalism and the Politics of the Labour Movement: Understanding the Political and Organisational Aspects of Labour Networking and Co-ordination," ed. Paul Stewart, *Employee Relations* 32, no. 6 (January 1, 2010): 538–56, available at https://doi.org/10.1108/01425451011083618.

67. Jamie K. McCallum, *Global Unions, Local Power: The New Spirit of Transnational Labor Organizing* (Ithaca, NY: Cornell University Press, 2013).

68. Fichter, "Building Union Power Across Borders."

69. Though the United States protects the right to organize in principle, violations are common and all but tolerated by an overburdened and ineffective labor board. Moreover, the United States has yet to ratify pertinent clauses within the ILO, and the Southeast is the most union-hostile region in the country.

70. Sarah Ashwin et al., "Spillover Effects across Transnational Industrial Relations Agreements: The Potential and Limits of Collective Action in Global Supply Chains," *ILR Review* 73, no. 4 (August 1, 2020): 1015, available at https://doi.org/10.1177/0019793919896570.

71. Arturo Rosenblueth, Norbert Wiener, and Julian Bigelow, "Behavior, Purpose and Teleology," *Philosophy of Science* 10, no. 1 (1943): 18–24.

72. Michael Fichter et al., "The Transformation of Organised Labour: Mobilising Power Resources to Confront 21st Century Capitalism," *Friedrich Ebert Stiftung*, July 1, 2018, available at https://library.fes.de/pdf-files/iez/14589.pdf; Whittall et al., "Workplace Trade Union Engagement."

73. Ashwin et al., "Spillover Effects across Transnational Industrial Relations Agreements."

74. Hammer and Riisgaard, "Labour and Segmentation in Value Chains."

75. Veronika Dehnen, "Transnational Alliances for Negotiating International Framework Agreements: Power Relations and Bargaining Processes between Global Union Federations and European Works Councils," *British Journal of Industrial Relations* 51, no. 3 (2013): 577–600, available at https://doi.org/10.1111/bjir.12038; Veronika Dehnen and Ludger Pries, "International Framework Agreements: A Thread in the Web of Transnational Labour Regulation," *European Journal of Industrial Relations* 20, no. 4 (December 1, 2014): 335–50, available at https://doi.org/10.1177/0959680113519187.

76. Aranea, González Begega, and Köhler, "European Works Council as a Management Tool."

77. Ann Cecilie Bergene, "Trade Unions Walking the Tightrope in Defending Workers' Interests: Wielding a Weapon Too Strong?" *Labor Studies Journal* 32, no. 2 (June 1, 2007): 142–66, available at https://doi.org/10.1177/0160449X07299703; Jeroen Merk, "Jumping Scale and Bridging Space in the Era of Corporate Social Responsibility: Cross-Border Labour Struggles in the Global Garment Industry," *Third World Quarterly* 30, no. 3 (2009): 599.

78. To be clear, this book is relatively agnostic as to whether the company or the industry is the appropriate basis for negotiation. It is clear, however, that companies may pit scales against one another, creating yet another basis for competitive pressure.

CHAPTER 5

1. Anonymous informant, interview with author, September 20, 2018.

2. George Rawick, "Working Class Self-Activity," May 1969, available at https://www.marxists.org/archive/rawick/1969/xx/self.html.

3. Jennifer Earl, "The Dynamics of Protest-Related Diffusion on the Web," *Information, Communication and Society* 13, no. 2 (March 1, 2010): 209–25, available at https://doi.org/10.1080/13691180902934170.

4. Ion Vasi, David Strang, and Arnout van de Rijt, "Tea and Sympathy: The Tea Party Movement and Republican Precommitment to Radical Conservatism in the 2011 Debt-Limit Crisis," *Mobilization: An International Quarterly* 19, no. 1 (April 2, 2014): 1–22, available at https://doi.org/10.17813/maiq.19.1.w30q317v25r35603.

5. Manuel Castells, *Networks of Outrage and Hope: Social Movements in the Internet Age* (Cambridge, UK: John Wiley and Sons, 2013).

6. Eric Blanc, *Red State Revolt: The Teachers' Strike Wave and Working-Class Politics* (London: Verso, 2019).

7. Nathan Lillie and Miguel Martínez Lucio, "Rollerball and the Spirit of Capitalism: Competitive Dynamics within the Global Context, the Challenge to Labour Transnationalism, and the Emergence of Ironic Outcomes," *Critical Perspectives on International Business* 8, no. 1 (January 1, 2012): 82, available at https://doi.org/10.1108/17422041211197576.

8. Malcolm Gladwell, "Small Change," *New Yorker*, September 27, 2010, available at https://www.newyorker.com/magazine/2010/10/04/small-change-malcolm-gladwell; Kevin Lewis, Kurt Gray, and Jens Meierhenrich, "The Structure of Online Activism," *Sociological Science* 1 (February 18, 2014): 1–9, available at https://doi.org/10.15195/v1.a1.

9. Rick Fantasia, *Cultures of Solidarity: Consciousness, Action, and Contemporary American Workers* (Berkeley: University of California Press, 1989).

10. Jack Ewing, "Volkswagen Reaches Deal in U.S. over Emissions Scandal," *New York Times*, April 21, 2016, sec. Business, available at https://www.nytimes.com/2016/04/22/business/international/volkswagen-emissions-settlement.html; Barry Hirsch, "What You Need to Know about the VW Emissions Scandal," *Los Angeles Times*, October 7, 2015, available at https://www.latimes.com/business/autos/la-fi-hy-volkswagen-qa-html-20151007-htmlstory.html.

11. Scott Neuman, "Volkswagen Names New CEO amid Emissions-Testing Scandal," NPR.org, September 25, 2015, available at https://www.npr.org/sections/thetwo-way/2015/09/25/443453460/vw-names-new-ceo-amid-emission-testing-scandal.

12. Hirsch, "What You Need to Know about the VW Emissions Scandal."

13. Mike Pare, "Glover Visits VW Plant, Union Members—VW Global Sales Fall in October amid Emissions Scandal," *CTFP*, November 14, 2015.

14. Multinationals concentrate key decision-making authority at a single corporate headquarters, which serves the center of global operations, while transnationals delegate authority to satellite locations, even as headquarters retains ultimate control. Lorraine Eden and Evan H. Potter, eds., *Multinationals in the Global Political Economy* (New York: St. Martin's, 1993). Even as other large multinationals shifted toward transnational-style organization, VW was something of a holdout, retaining the primacy of its Wolfsburg headquarters well into the twenty-first century.

15. Gabe Nelson, "VW's Tenn. Plant Chief Is Reassigned," *Automotive News: Detroit* 88, no. 6614 (March 31, 2014): 27.

16. Ryan Beene, "VW's Lost Year? Automaker's Efforts to Clear Diesel Cloud Sputter," *Automotive News* 90, no. 6708 (January 18, 2016): 3.

17. Dave Flessner, "UAW Advances Plan for Works Council in Chattanooga," *CTFP*, May 8, 2015.

18. Mike Pare, "UAW Seeks 'Fresh Start'—Opponents at Chattanooga VW Plant Call Local 42 'a Show,'" *CTFP*, July 11, 2014.

19. Pare, "UAW Seeks 'Fresh Start.'"

20. Daniel Gross et al., *Solidarity Unionism at Starbucks*, PM Press Pamphlet Series (Oakland, CA: PM Press, 2011), available at http://site.ebrary.com/id/10440664.

21. The experience of the Amazon Labor Union in Staten Island represents an intriguing counterexample in which the upstart union achieved a notable victory despite its status as an outsider to the House of Labor.

22. Charles J. Morris, *The Blue Eagle at Work: Reclaiming Democratic Rights in the American Workplace* (Ithaca, NY: Cornell University Press, 2005).

23. Dorothy Sue Cobble, *Lost Ways of Unionism: Historical Perspectives on Reinventing the Labor Movement* (Ithaca: Cornell University Press, 2001); Dorothy Sue Cobble, *Craft Unionism Revisited: The Case of the Waitress Locals* (New Brunswick, NJ: Rutgers Institute of Management and Labor Relations, 1989); Dorothy Sue Cobble, "Rethinking Troubled Relations between Women and Unions: Craft Unionism and Female Activism," *Feminist Studies* 16, no. 3 (1990): 519–48, available at https://doi.org/10.2307/3178018; Dorothy Sue Cobble, "Organizing the Postindustrial Work Force: Lessons from the History of Waitress Unionism," *ILR Review*, June 25, 2016, available at https://doi.org/10.1177/001979399104400302.

24. Chris Brooks, "Organizing Volkswagen: A Critical Assessment," *WorkingUSA* 19 (September 1, 2016), available at https://doi.org/10.1111/wusa.12249.

25. Stanley Aronowitz, *The Death and Life of American Labor: Toward a New Worker's Movement* (London: Verso, 2014).

26. Mike Pare, "New UAW Local President to Focus on VW Recognition," *CTFP*, October 5, 2014.

27. Mike Pare, "Skilled Workers Vote at VW on UAW," *CTFP*, December 3, 2015.

28. Mike Pare, "UAW Seeks New Vote at VW Plant," *CTFP*, October 24, 2015.

29. Mike Pare, "VW Objects to Vote by One Group of Workers," *CTFP*, November 3, 2015.

30. Mike Pare, "VW, Union Spar over Workers in Voting Bloc," *CTFP*, November 4, 2015.

31. Mike Pare, "UAW Gets OK to Hold Union Vote," *CTFP*, November 19, 2015.

32. Brooks, "Organizing Volkswagen."

33. Mike Pare, "'Historic'—UAW Wins at Volkswagen—Skilled Trades Workers Vote to Unionize in Chattanooga," *CTFP*, December 5, 2015.

34. Ted Evanoff, "UAW Continues Pressure on VW Chattanooga," *CTFP*, April 28, 2017.

35. Andy Sher and Mike Pare, "Haslam: 'Timing Isn't Great' for United Auto Workers Election at Volkswagen," *CTFP*, October 28, 2015.

36. Sher and Pare, "Haslam."

37. Mike Pare, "VW Appeals UAW Election in Chattanooga," *CTFP*, December 31, 2015.

38. Mike Pare, "UAW Local 42 Files Charge against VW over Chattanooga Plant," *CTFP*, December 21, 2015.

39. Mike Pare, "NLRB Rejects Challenge from VW—Automaker Weighs Next Move after UAW Wins Ruling," *CTFP*, April 14, 2016.

40. Mike Pare, "NLRB Files Unfair Labor Practices against VW," *CTFP*, April 28, 2016.

41. Mike Pare, "Labor Board Ruling Could Affect VW Chattanooga Facility," *CTFP*, December 19, 2017.

42. Mike Pare, "VW, UAW Quarrel over Local Plant Vote," *CTFP*, April 17, 2019.

43. Chris Brooks, "A Tumultuous Week for Chattanooga Volkswagen Workers," *Labor Notes*, May 24, 2019, available at https://www.labornotes.org/2019/05/tumultuous-week-chattanooga-volkswagen-workers.

44. Mike Pare, "Lawyer for VW Wants to Quash New Election—Union Calls Company's Tactics 'Disgraceful,'" *CTFP*, April 26, 2019.

45. Mike Pare, "UAW Wants to Disclaim VW Chattanooga Win," *CTFP*, April 16, 2019.

46. Mike Pare, "Board Ruling Puts UAW Union Election on Hold," *CTFP*, May 4, 2019; Mike Pare, "Union Renews Call for VW Chattanooga Election," *CTFP*, May 7, 2019.

47. Mike Pare, "Labor Board Certifies VW Union Vote," *CTFP*, June 29, 2019.

48. "UAW 2024 National Cap Conference," January 24, 2024, available at https://www.facebook.com/watch/live/?ref=watch_permalink&v=1418813415378995.

49. Chris Brooks, "Tennessee Governor Leads Anti-union Captive Audience Meeting at VW," *Labor Notes*, April 29, 2019, available at https://labornotes.org/blogs/2019/04/tennessee-governor-leads-anti-union-captive-audience-meeting-vw.

50. Mike Pare, "Anti-UAW Group Forms Ahead of Union Vote," *CTFP*, May 1, 2019.

51. Mike Pare, "VW Eyes UAW Recognition Here, Group Says," *CTFP*, April 8, 2014.

52. Dave Flessner, "Workers Claim VW Illegally Helped United Auto Workers," *CTFP*, March 14, 2014.

53. Mike Pare, "Group Hits German Union as Volkswagen Vote Nears," *CTFP*, November 24, 2015.

54. *Community Organization Engagement* (Chattanooga, TN: Volkswagen Group of America Chattanooga Operations, 2014), available at https://www.scribd.com/document/246408396/VW-Engagement-Policy.

55. Mike Pare, "UAW Gains Ground at Volkswagen—Union to Start Regular Meetings with Officials at Chattanooga Plant," *CTFP*, December 9, 2014.

56. Mike Pare, "UAW Questions VW Labor Rival over Funding—ACE Says UAW Is 'Detroit-Funded,'" *CTFP*, September 5, 2015.

57. Mike Pare, "What's Next? Future at VW Chattanooga Eyed amid Union Loss," *CTFP*, June 16, 2019; Mike Pare, "Counting Costs and Benefits of UAW at VW," *CTFP*, June 13, 2019. A plant spokesperson cited VW's bonus program as another impediment to unionization, noting that average team member bonuses for 2018 equaled $3,682. She declined to mention, however, that bonuses at UAW-represented legacy automakers like GM averaged $10,750. Pare, "Counting Costs and Benefits of UAW at VW."

58. Mike Pare, "Former CEO Set to Return to VW Plant—Fischer Will Replace Pinto as Chief Executive," *CTFP*, May 25, 2019.

59. Pare Mike, "Volkswagen Layoffs in Chattanooga Marked by Surprise, Regret," *CTFP*, April 19, 2013, available at https://www.timesfreepress.com/news/2013/apr/19/vw-layoffs-marked-by-surprise-regret/.

60. Pare, "What's Next?"

61. Mike Pare, "VW Creating 2,000 Jobs; UAW Gains Ground," *CTFP*, December 31, 2014.

62. Mike Pare and Dave Flessner, "Haslam, UAW Spar over Union Impact," *CTFP*, December 17, 2014.

63. Mike Pare, "The Ripple Effect—What Does the UAW Vote Mean for Volkswagen and Auto Manufacturing across the South?" *CTFP*, December 6, 2015.

64. Chris Brooks, "Volkswagen Declares War against Works Council and German Union," *Labor Notes*, June 12, 2019, available at https://labornotes.org/blogs/2019/06/volkswagen-declares-war-against-works-council-and-german-union.

65. Brooks, "Volkswagen Declares War."

66. Pare, "What's Next?"

67. Private Facebook group post.

68. Chris Brooks, "Volkswagen Reverses Its Union Neutrality, Workers Say," *Labor Notes*, October 28, 2015, available at http://labornotes.org/2015/10/volkswagen-reverses-its-union-neutrality-workers-say.

69. Chris Brooks, "Volkswagen Workers Celebrate Election Win, But Question Union's Partnership Strategy," *Labor Notes*, December 5, 2015, available at https://labor notes.org/2015/12/volkswagen-workers-celebrate-election-win-question-unions-part nership-strategy.

70. Amanda Aronczyk, "The UAW's Decade-Long Fight to Form a Union at VW's Chattanooga Plant (Update): Planet Money," NPR, April 24, 2024, available at https://www.npr.org/2024/04/24/1197958834/uaw-united-auto-workers-union-strike-volkswa gen-chattanooga-update.

71. Barry Eidlin and Jane Slaughter, "After Decades of Corrupt, Antidemocratic Rule, UAW Members Are Finally Electing Their Leaders," *Jacobin*, October 20, 2022, available at https://jacobin.com/2022/10/uaw-reform-leadership-elections-uawd.

72. Robert Combs, "Analysis: The Big 3–UAW Pay Raise of (TBD)% Will Make History," Bloomberg Law, September 22, 2023, available at https://news.bloomberglaw .com/bloomberg-law-analysis/analysis-the-big-3-uaw-pay-raise-of-tbd-will-make -history.

73. Mike Pare, "Counting the Costs, Benefits of UAW Membership at Volkswagen Chattanooga," *CTFP*, June 12, 2019, available at https://www.timesfreepress.com/news /2019/jun/12/costs-benefits-uaw-membership/.

74. On the Line, "UAW's Massive Campaign in the South," April 13, 2024, available at https://www.laborradionetwork.org/lrpn-members-1/on-the-line.

75. Luis Feliz Leon, "Twenty-Five Thousand Auto Workers Are Now on Strike at the Big 3," *Labor Notes*, September 29, 2023, available at https://labornotes.org/2023/09/twen ty-five-thousand-auto-workers-are-now-strike-big-3.

76. Luis Feliz Leon, "Auto Workers Direct Momentum toward Organizing Plants across the U.S.," *Labor Notes*, November 30, 2023, available at https://labornotes.org/2023/11 /auto-workers-direct-momentum-toward-organizing-plants-across-us.

77. Neal E. Boudette, "F.B.I. Raids U.A.W. Chief's Home as Financial Inquiry Widens," *New York Times*, August 28, 2019, sec. Business, available at https://www.nytimes.com /2019/08/28/business/uaw-fbi-raid.html; Noam Scheiber and Neal E. Boudette, "Behind a U.A.W. Crisis: Lavish Meals and Luxury Villas," *New York Times*, December 26, 2019, sec. Business, available at https://www.nytimes.com/2019/12/26/business/uaw-gary-jones -investigation.html; Noam Scheiber and Neal E. Boudette, "U.A.W. Corruption Case Widens as Former Chief Is Charged," *New York Times*, March 5, 2020, sec. Business, available at https://www.nytimes.com/2020/03/05/business/uaw-indictment.html.

78. Mike Pare, "Unionization Battle at VW Heating Up—Ads Charge Corruption by Automaker, UAW," *CTFP*, June 1, 2019.

79. Eidlin and Slaughter, "After Decades of Corrupt, Antidemocratic Rule."

80. Brooks, "Tennessee Governor Leads Anti-Union Captive Audience Meeting at VW."

81. Matt Patterson, "UAW Vows Attack on Chattanooga," *CTFP*, June 17, 2014.

82. However, this time around the union could not access the shop floor, so its contact with workers was minimal.

83. Brooks, "Tennessee Governor Leads Anti-Union Captive Audience Meeting at VW."

84. Wikipedia, s.v. "*Fortune* Global 500," last modified July 19, 2024, available at https:// en.wikipedia.org/w/index.php?title=Fortune_Global_500&oldid=1235563047.

85. Mike Pare, "Pro-UAW VW Workers Cite Safety, Other Concerns," *CTFP*, June 9, 2019.

86. Chris Brooks, "Why the UAW Lost Again in Chattanooga," *Labor Notes*, June 14, 2019, available at https://labornotes.org/2019/06/why-uaw-lost-again-chattanooga.

CHAPTER 6

1. Joseph Schumpeter, *Capitalism, Socialism, and Democracy: Third Edition* (New York: Harper Perennial Modern Classics, 2008).

2. Marshall Ganz, "Resources and Resourcefulness: Strategic Capacity in the Unionization of California Agriculture, 1959–1966," *American Journal of Sociology* 105, no. 4 (January 2000): 1012, available at https://doi.org/10.1086/210398.

3. E. J. Langer and L. G. Imber, "When Practice Makes Imperfect: Debilitating Effects of Overlearning," *Journal of Personality and Social Psychology* 37, no. 11 (November 1979): 2014–24, available at https://doi.org/10.1037//0022-3514.37.11.2014.

4. Kenneth O. McGraw and John C. McCullers, "Evidence of a Detrimental Effect of Extrinsic Incentives on Breaking a Mental Set," *Journal of Experimental Social Psychology* 15, no. 3 (May 1, 1979): 285–94, available at https://doi.org/10.1016/0022-1031(79)90 039-8.

5. Connie J. Gersick, "Revolutionary Change Theories: A Multilevel Exploration of the Punctuated Equilibrium Paradigm," *Academy of Management Review* 16, no. 1 (1991): 10–36, available at https://doi.org/10.2307/258605.

6. Judith Stepan-Norris and Maurice Zeitlin, *Left Out: Reds and America's Industrial Unions* (Cambridge: Cambridge University Press, 2002), 196.

7. Dan Georgakas, Marvin Surkin, and Manning Marable, *Detroit: I Do Mind Dying; A Study in Urban Revolution*, rev. ed. (Cambridge, MA: South End, 1999).

8. Diane Feeley, "Organizing Auto in the South," *Solidarity* (blog), May 21, 2024, available at https://solidarity-us.org/organizing-auto-in-the-south/.

9. Alexandra Bradbury, "Chrysler Workers Vote 2 to 1 to Reject Two-Tier Pact," *Labor Notes*, October 1, 2015, available at https://labornotes.org/2015/09/so-far-chrysler-work ers-roundly-rejecting-two-tier-pact.

10. Micah Uetricht and Barry Eidlin, "U.S. Union Revitalization and the Missing 'Militant Minority,'" *Labor Studies Journal* 44, no. 1 (March 1, 2019): 36–59, available at https://doi.org/10.1177/0160449X19828470.

11. Barry Eidlin and Jane Slaughter, "After Decades of Corrupt, Antidemocratic Rule, UAW Members Are Finally Electing Their Leaders," *Jacobin*, October 20, 2022, available at https://jacobin.com/2022/10/uaw-reform-leadership-elections-uawd.

12. Luis Feliz Leon and Jane Slaughter, "It's a New Day in the United Auto Workers," *Labor Notes*, March 17, 2023, available at https://labornotes.org/2023/03/its-new-day-unit ed-auto-workers.

13. Jonah Furman, "Auto Workers Win Direct Democracy in Referendum," *Labor Notes*, December 1, 2021, available at https://labornotes.org/2021/12/auto-workers-win-direct -democracy-referendum.

14. Jane Slaughter, "Challengers Win Big in UAW Elections; Presidency Headed to Run-Off," *Labor Notes*, December 2, 2022, available at https://labornotes.org/2022/12 /challengers-win-big-uaw-elections-presidency-headed-run.

15. Dan DiMaggio and Keith Brown, "Auto Workers Have Big Demands for the Big 3," *Labor Notes*, August 17, 2023, available at https://labornotes.org/2023/08/auto-work ers-have-big-demands-big-3.

16. Keith Brown, "Auto Workers Debate Contracts: Tall Gains, Taller Expectations," *Labor Notes*, November 10, 2023, available at https://labornotes.org/2023/11/auto-work ers-debate-contracts-tall-gains-taller-expectations.

17. Ganz, "Resources and Resourcefulness," 1015.

18. Ganz, "Resources and Resourcefulness," 1017.

19. Mike Pare, "Biden Lauds UAW Win at VW Chattanooga," *CTFP*, April 20, 2024, available at https://www.timesfreepress.com/news/2024/apr/20/biden-lauds-uaw-win -at-vw-chattanooga/.

20. Luis Feliz Leon, "Auto Workers Spare Big 3, Win Landmark Just Transition at General Motors," *Labor Notes*, October 6, 2023, available at https://labornotes.org/2023 /10/auto-workers-spare-big-3-win-landmark-just-transition-general-motors.

21. Luis Feliz Leon, "Auto Workers Halt Stellantis's Biggest Moneymaker," *Labor Notes*, October 23, 2023, available at https://labornotes.org/2023/10/auto-workers-halt -stellantis-biggest-moneymaker.

22. Leon, "Auto Workers Halt Stellantis's Biggest Moneymaker."

23. Jake Alimahomed-Wilson and Immanuel Ness, eds., *Choke Points: Logistics Workers Disrupting the Global Supply Chain* (London: Pluto, 2018).

24. Dale Buss, "GM's Aftermarket Parts Sales Rise; So Does Warehouse Investment," *Forbes*, May 1, 2023, available at https://www.forbes.com/sites/dalebuss/2023/04/30/gms -aftermarket-parts-sales-rise-so-does-warehouse-investment/.

25. Michael Wayland, "Stellantis Could Close 18 Facilities under UAW Deal—Here Are the Full Details of Its Latest Offer," CNBC, September 18, 2023, available at https:// www.cnbc.com/2023/09/18/uaw-strike-stellantis-could-close-18-facilities-under-deal .html.

26. Though no replacement workers were used at assembly plants, newly hired temps were brought in at a parts facility in Swartz Creek, Michigan. See Jane Slaughter, "Scabs Deployed at GM Parts Distribution Centers," *Labor Notes*, September 25, 2023, available at https://labornotes.org/2023/09/scabs-deployed-gm-parts-distribution -centers.

27. Jane Slaughter, "Ford and GM Agree to End at Least One Tier, Stellantis Still Holding Out," *Labor Notes*, September 25, 2023, available at https://labornotes.org/blogs/2023 /09/ford-and-gm-agree-end-least-one-tier-stellantis-still-holding-out.

28. Dan DiMaggio, "Big 3 Buckled as Stand-Up Strike Spread," *Labor Notes*, October 31, 2023, available at https://labornotes.org/2023/10/big-3-buckled-stand-strike -spread.

29. Lisa Xu, "UAW Supporters to Hit the Pavement at Dealerships," *Labor Notes*, September 28, 2023, available at https://www.labornotes.org/2023/09/uaw-supporters -hit-pavement-dealerships.

30. Dan DiMaggio and Courtney Smith, "Auto Workers Kick Off Bargaining with 'Members' Handshake,'" *Labor Notes*, July 20, 2023, available at https://labornotes.org /2023/07/auto-workers-kick-bargaining-members-handshake.

31. Temps falling short of the ninety-day or nine-month benchmark could still be dismissed without cause.

32. Leon, "Auto Workers Spare Big 3."

33. Luis Feliz Leon and Lisa Xu, "Auto Workers Strike Spreads to 38 Parts Depots," *Labor Notes*, September 22, 2023, available at https://labornotes.org/2023/09/auto-workers -strike-spreads-38-parts-depots.

34. Edward McClelland, *Midnight in Vehicle City: General Motors, Flint, and the Strike That Created the Middle Class* (Beacon Press, 2021).

35. Kate Bronfenbrenner et al., *Organizing to Win: New Research on Union Strategies* (Ithaca, NY: Cornell University Press, 1998).

36. Seymour Martin Lipset, *Union Democracy; the Internal Politics of the International Typographical Union* (Glencoe, IL: Free Press, 1956).

37. Judith Stepan-Norris and Maurice Zeitlin, "Insurgency, Radicalism, and Democracy in America's Industrial Unions," *Social Forces* 75, no. 1 (1996): 1–32, available at https://doi.org/10.2307/2580755.

38. Stepan-Norris and Zeitlin, "Insurgency, Radicalism, and Democracy."

39. Lipset, *Union Democracy*, 13.

40. Robert Michels, *Political Parties: A Sociological Study of the Oligarchical Tendencies of Modern Democracy* (London: Routledge, 2017).

41. Michels, *Political Parties*.

42. Keith Brown and Jane Slaughter, "Auto Workers Convention Lurches towards Reversing Concessions," *Labor Notes*, March 30, 2023, available at https://labornotes.org/2023/03/auto-workers-convention-lurches-towards-reversing-concessions.

43. Keith Brown, "Slow Walks and Tough Talk: Auto Workers Turn the Screws," *Labor Notes*, September 28, 2023, available at https://labornotes.org/2023/09/slow-walks-and-tough-talk-auto-workers-turn-screws.

44. Phoebe Wall Howard, "UAW Members Reject Tentative Agreement with Mack Trucks, Prep for Strike Monday," *Detroit Free Press*, October 10, 2023, available at https://www.freep.com/story/money/cars/2023/10/08/uaw-members-reject-tentative-agreement-with-mack-trucks-to-strike/71115000007/.

45. Stepan-Norris and Zeitlin, *Left Out*, 85.

46. Nicholas Thoburn, *Deleuze, Marx and Politics* (London: Routledge, 2003), 34.

47. Barbara Ellen Smith, *Digging Our Own Graves: Coal Miners and the Struggle over Black Lung Disease* (Chicago: Haymarket Books, 2020), 170, available at http://ebookcentral.proquest.com/lib/bmcc/detail.action?docID=6184204.

48. Mike Parker and Martha Gruelle, *Democracy Is Power: Rebuilding Unions from the Bottom Up* (Detroit: Labor Notes, 1999), 154.

49. *Pittsburgh Post-Gazette*, "MFD Should Know Better," July 10, 1973.

50. Aaron Brenner, Robert Brenner, and Cal Winslow, eds., *Rebel Rank and File: Labor Militancy and Revolt from Below during the Long 1970s* (London: Verso Books, 2010).

51. Parker and Gruelle, *Democracy Is Power*.

52. Thaddeus Russell, *Out of the Jungle: Jimmy Hoffa and the Remaking of the American Working Class* (Philadelphia, PA: Temple University Press, 2003).

53. Marshall Ganz, *Why David Sometimes Wins: Leadership, Organization, and Strategy in the California Farm Worker Movement*, reprint ed. (Oxford: Oxford University Press, 2010), 239–54.

54. Ruth Milkman, *L.A. Story: Immigrant Workers and the Future of the U.S. Labor Movement* (New York: Russell Sage Foundation, 2006), 153.

55. Nelson Lichtenstein, *Walter Reuther: The Most Dangerous Man in Detroit* (Urbana: University of Illinois Press, 1997), 133.

CHAPTER 7

1. Mike Pare, "UAW Complains about VW in Chattanooga in National Labor Relations Board Filing," *CTFP*, December 3, 2013, available at https://www.timesfreepress.com/news/2023/dec/11/uaw-complains-about-vw-in-chattanooga-in-nlrb/.

2. "Exclusive: UAW President Shawn Fain on How the Auto Workers Won and What's Next," *In These Times*, November 7, 2023, available at https://inthesetimes.com/article/exclusive-interview-uaw-president-shawn-fain.

3. Management noted that base wages had increased by 56 percent over the ten years of the plant's operations, significantly outpacing inflation (31.7 percent).

4. Reuters, "Volkswagen Becomes the Latest Automaker to Hike Wages for U.S. Factory Workers," November 22, 2023, sec. Autos and Transportation, available at https://www.reuters.com/business/autos-transportation/volkswagen-becomes-latest-automaker-hike-wages-us-factory-workers-2023-11-22/.

5. Among the first wave of transplants, pay was less of an issue. Though shop-floor conditions in early Japanese transplants broke sharply from the Fordist model, pay and benefits were initially set close to UAW levels, a strategy to avert unionization.

6. Office of the Governor, "Gov. Lee Joins Coalition of Governors in Opposing UAW's Unionization Campaign," April 16, 2024, available at https://www.tn.gov/governor/news/2024/4/16/gov--lee-joins-coalition-of-governors-in-opposing-uaw-s-unionization-campaign.html.

7. "Still No to UAW," available at http://no2uaw.com, accessed September 14, 2024.

8. Juliana Kim, "Black and Latina Women Helped Propel Gains for Unions in 2023, Finds a New Study," NPR, February 5, 2024, sec. National, available at https://www.npr.org/2024/02/05/1228933397/union-membership-black-and-latina-women-2023.

9. Jamie L. LaReau, "UAW Scores Historic Landslide Victory to Unionize the First Foreign Automaker in the US," *Detroit Free Press*, available at https://www.freep.com/story/money/cars/2024/04/19/results-uaw-union-vote-volkswagen-chattanooga-tennessee/73374618007/.

10. LaReau, "UAW Scores Historic Landslide Victory."

11. Gustavo Henrique Ruffo, "Volkswagen Starts ID.4 Production in Chattanooga with SK Innovation Cells Made in Georgia," Auto Evolution, July 26, 2022, available at https://www.autoevolution.com/news/volkswagen-starts-id4-production-in-chattanooga-with-sk-innovation-cells-made-in-georgia-194522.html.

12. *MotorWeek*, "Volkswagen ID.4 Now Assembled at Chattanooga Plant," July 26, 2022, available at https://motorweek.org/this_just_in/volkswagen-id-4-now-assembled-at-chattanooga-plant/.

13. David Dayen, "New Clean-Car Rule May Transform Unions in America," *American Prospect*, March 22, 2024, available at https://prospect.org/api/content/3334e9c4-e7bc-11ee-beef-12163087a831/.

14. Luis Feliz Leon, "Southern Autoworkers Organize, Business Class Tries to Wallop Them," *American Prospect*, February 15, 2024, available at https://prospect.org/api/content/94413448-cb51-11ee-900a-12163087a831/.

15. Dayen, "New Clean-Car Rule May Transform Unions in America."

16. Mike Pare, "VW Chattanooga Expects Assembly to Grow by a Third in 2024," *CTFP*, February 16, 2024, available at https://www.timesfreepress.com/news/2024/feb/16/vw-chattanooga-expects-assembly-to-grow-by-a/.

17. Mike Pare, "Volkswagen's New Battery-Powered 'Microbus' Will Join the Chattanooga-Made ID.4 SUV in Dealerships," Yahoo News, March 10, 2022, available at https://www.yahoo.com/news/volkswagens-battery-powered-microbus-join-050100297.html.

18. Mike Pare, "VW Chattanooga Plant Launches Biggest Hiring Surge since Plant Opening," *CTFP*, April 8, 2022, available at https://www.timesfreepress.com/news/2022/apr/08/vw-chattanooga-launches-biggest-hiring-surge/.

19. Office of the Governor, "Gov. Lee Joins Coalition of Governors in Opposing UAW's Unionization Campaign," April 16, 2024, available at https://www.tn.gov/governor/news/2024/4/16/gov--lee-joins-coalition-of-governors-in-opposing-uaw-s-unionization-campaign.html.

20. Donatella Della Porta and Martín Portos, "Social Movements in Times of Inequalities: Struggling against Austerity in Europe," *Structural Change and Economic Dynamics* 53 (June 1, 2020): 116–26, available at https://doi.org/10.1016/j.strueco.2020.01.011.

21. Chantal Pezold, Simon Jäger, and Patrick Nüss, "Labor Market Tightness and Union Activity" (Working Paper Series, National Bureau of Economic Research, December 2023), available at https://doi.org/10.3386/w31988.

22. Marshall Ganz, "Resources and Resourcefulness: Strategic Capacity in the Unionization of California Agriculture, 1959–1966," *American Journal of Sociology* 105, no. 4 (January 2000): 1014–18, available at https://doi.org/10.1086/210398.

23. Even with the recently announced increase in the organizing budget, organizing remains a much less significant portion of the UAW's operating costs than representation and services for existing members.

24. Ganz, "Resources and Resourcefulness."

25. Ganz, "Resources and Resourcefulness," 1014.

26. Luis Feliz Leon, "Auto Workers Direct Momentum toward Organizing Plants across the U.S.," *Labor Notes*, November 30, 2023, available at https://labornotes.org/2023/11/auto-workers-direct-momentum-toward-organizing-plants-across-us.

27. Steven Greenhouse, "Volkswagen 'the First Domino to Fall' after Union Vote, Says UAW President," *The Guardian*, April 22, 2024, sec. US News, available at https://www.theguardian.com/us-news/2024/apr/22/united-auto-workers-auw-shawn-fain.

28. More Perfect Union [@MorePerfectUS], "Right after the @UAW Won at the Big 3, Toyota gave raises at their U.S. plants. But like union President Shawn Fain says: 'Toyota isn't giving out raises out of the goodness of their heart. They did it now because the company knows we're coming for them.' https://t.co/tawN002IZz," Twitter, November 3, 2023, available at https://x.com/MorePerfectUS/status/1720443876350046524.

29. Amanda Aronczyk, "The UAW's Decade-Long Fight to Form a Union at VW's Chattanooga Plant (Update): Planet Money," NPR, April 24, 2024, available at https://www.npr.org/2024/04/24/1197958834/uaw-united-auto-workers-union-strike-volkswagen-chattanooga-update.

30. Annette D. Bernhardt and Labor and Employment Relations Association, eds., *The Gloves-Off Economy: Workplace Standards at the Bottom of America's Labor Market*, Labor and Employment Relations Association Series (Champaign: Labor and Employment Relations Association, University of Illinois at Urbana-Champaign, 2008).

31. Steve Greenhouse, "'Victories Would Be Nothing Less Than an Earthquake': Can UAW Win in the South?" *The Guardian*, April 17, 2024, available at https://portside.org/2024-04-17/victories-would-be-nothing-less-earthquake-can-uaw-win-south.

32. Luis Feliz Leon, "Tennessee Volkswagen Workers Have Filed for a Union Election," *Jacobin*, March 18, 2024, available at https://jacobin.com/2024/03/tennessee-volkswagen-uaw-union-election.

33. Leon, "Auto Workers Direct Momentum."

34. *Jacobin Radio*, "UAW's Southern Campaign w/ Chattanooga Auto Workers," May 23, 2024, available at https://podcasts.apple.com/us/podcast/jacobin-radio-uaws-southern-campaign-w-chattanooga/id791564318?i=1000656516741.

35. Chris Isidore, "UAW Union Vote at Volkswagen Could Have Big Implications for US Auto Industry and Labor's Strength in South," CNN, April 19, 2024, available at https://www.cnn.com/2024/04/19/business/uaw-volkswagen-vote/index.html.

36. Breana Noble and Kalea Hall, "Look Inside the UAW's Work to Organize Transplants, EV Autoworkers," *The Detroit News*, December 11, 2023, available at https://www

.detroitnews.com/story/business/autos/foreign/2023/12/11/uaw-organize-transplants-ev-autoworkers/71678033007/.

37. Mike Elk, "After the Pandemic and the Stand Up Strike, Everything Changed at Volkswagen in Chattanooga," Payday Report, March 27, 2024, available at https://paydayreport.com/after-the-pandemic-and-the-stand-up-strike-everything-changed-at-volkswagen-in-chattanooga/.

38. Breana Noble, "UAW Expands GM Strike to Tennessee Plant," *Detroit News*, October 28, 2023, available at https://www.detroitnews.com/story/business/autos/2023/10/28/uaw-gm-spring-hill-assembly-plant-tennessee-strike/71365399007/.

39. UAW, *Stand Up VW*, YouTube, December 7, 2023, available at https://www.youtube.com/watch?v=l8cfWEeLOCc.

40. UAW, *The UAW Difference at Volkswagen* (Detroit: UAW, 2024), available at https://uaw.org/wp-content/uploads/2024/04/UAW-Difference-at-VW.pdf.

41. UAW, *Improving Work-Life Balance at VW* (Detroit: UAW, 2024), available at https://uaw.org/wp-content/uploads/2024/02/IMPROVING-WORK-LIFE-BALANCE-AT-VOLKSWAGEN.pdf.

42. Pare, "VW Chattanooga Plant Launches Biggest Hiring Surge."

43. Joan McClane, "The German Way: After Battling a Decade to Join the UAW, Chattanooga Auto Workers Shook the South and Finally Secured Success," *CTFP*, June 2, 2024, available at https://www.timesfreepress.com/news/2024/jun/02/the-german-way-after-battling-a-decade-to-join-the-uaw-chattanooga-auto-workers-shook-the-south-and-finally-secured-success/.

44. Local 3 News Staff, "Update: Six State Governors Form Coalition to Oppose UAW's Efforts as VW Union Vote Approaches," Local3News.com, April 16, 2024, available at https://www.local3news.com/local-news/update-six-state-governors-form-coalition-to-oppose-uaw-s-efforts-as-vw-union-vote/article_9f0fbe8a-f04a-11ee-848a-cb977a198af0.html.

45. Mike Pare, "VW Chattanooga, Area Employers Face New Labor Future," *CTFP*, April 20, 2024, available at https://www.timesfreepress.com/news/2024/apr/20/vw-chattanooga-area-employers-face-new-labor/.

46. *Jacobin Radio*, "UAW's Southern Campaign."

47. Even the 2023 contract, which took aim at the wage gap, did not completely bring the second tier up to parity. The UAW preemptively proclaimed the "elimination" of two-tier, but this victory extended only to wages.

48. Laurie Graham, *On the Line at Subaru-Isuzu: The Japanese Model and the American Worker* (Ithaca, NY: Cornell University Press, 1995).

49. Anonymous informant, interview with author, September 21, 2016.

50. U.S. Census Bureau, "QuickFacts: Chattanooga City, Tennessee," accessed July 12, 2024, available at https://www.census.gov/quickfacts/fact/table/chattanoogacitytennessee/PST045223#qf-headnote-a.

51. Pew Research Center, "Mixed Views of Impact of Long-Term Decline in Union Membership," *Pew Research Center* (blog), April 27, 2015, available at https://www.pewresearch.org/politics/2015/04/27/mixed-views-of-impact-of-long-term-decline-in-union-membership/.

52. Mike Elk, "After Ten-Year Battle, a Younger Generation Leads the Way at Volkswagen," *American Prospect*, April 19, 2024, available at https://prospect.org/api/content/a07da780-fdc6-11ee-9def-12163087a831/.

53. Ted Van Green, "Majorities of Adults See Decline of Union Membership as Bad for the U.S. and Working People," *Pew Research Center* (blog), March 12, 2024, available

at https://www.pewresearch.org/short-reads/2024/03/12/majorities-of-adults-see-dec
line-of-union-membership-as-bad-for-the-us-and-working-people/.

54. Leon, "Tennessee Volkswagen Workers Have Filed for a Union Election."

55. Kim, "Black and Latina Women Helped Propel Gains for Unions in 2023."

56. Troy Hunt, "Opinion: I've Worked at the VW Plant for More than a Decade. Here's Why We Need a Union," *CTFP*, February 22, 2024, available at https://www .timesfreepress.com/news/2024/feb/22/opinion-ive-worked-at-the-vw-plant-for-more -than/.

57. Courtney Vinopal, "Why Volkswagen Is Introducing Family-Friendly Policies for Shift Workers," *HR Brew*, September 10, 2025, available at https://www.hr-brew.com /stories/2024/05/10/volkswagen-family-friendly-policies.

58. At the time of the election, workers received between 96 and 144 hours of PTO a year (about twelve to eighteen full-time days), but when VW closed the plant for a week or two during the year for maintenance, workers had to use their PTO if they wanted to be paid for those days. Four planned shutdown days were listed as "company scheduled PTO," meaning the holidays counted against workers' PTO allowance. New hires received only twelve days of annual PTO, so these required "holidays" cut into their allocation significantly.

59. Local 42, "Protect Our PTO—Volkswagen Workers Stand Up!" July 11, 2022, available at https://local42.org/articles/article-one/.

60. UAW, *VW Chattanooga Ready To Vote YES*, YouTube, March 18, 2024, available at https://www.youtube.com/watch?v=Atm-Jqr9Klg.

61. Leon, "Tennessee Volkswagen Workers Have Filed for a Union Election."

62. Hunt, "Opinion."

63. Signaling the shift in priorities, several of the more high-profile labor struggles of the 1990s and 2000s were over the right to work *additional* overtime.

64. Shawn Fain, "Workers Deserve More Time for Themselves," *Jacobin*, March 15, 2024, available at https://jacobin.com/2024/03/shawn-fain-thirty-two-hour-workweek -speech.

65. Dan DiMaggio, "Auto Workers Call on Unions to Align Contract Expirations," *Labor Notes*, November 22, 2023, available at https://labornotes.org/blogs/2023/11/auto -workers-call-unions-align-contract-expirations.

66. Jasper Bernes, "Logistics, Counterlogistics and the Communist Prospect by Jasper Bernes," Endnotes 3, September 2013, https://endnotes.org.uk/articles/logistics -counterlogistics-and-the-communist-prospect.

67. AFL-CIO, "At What Cost? Workers Report Concerns about Volkswagen's New Manufacturing Jobs in Tennessee," May 2017, available at https://www.industriall-union.org /sites/default/files/uploads/documents/2017/SWITZERLAND/at_what_cost.pdf.

68. Hunt, "Opinion."

69. Alex Press, "Chattanooga VW Worker: 'This Will Change What People Think Is Possible,'" *Jacobin*, April 22, 2024, available at https://jacobin.com/2024/04/chattanoo ga-vw-uaw-unionization.

70. Luis Feliz Leon, "The UAW Is Gearing Up for Two Union Elections in the South," *Jacobin*, April 8, 2024, available at https://jacobin.com/2024/04/uaw-mercedes-volkswa gen-alabama-tennessee-union-election.

71. Keith Brown, "Auto Workers Debate Contracts: Tall Gains, Taller Expectations," *Labor Notes*, November 10, 2023, available at https://labornotes.org/2023/11/auto-workers -debate-contracts-tall-gains-taller-expectations.

72. Leon, "UAW Is Gearing Up for Two Union Elections in the South."

73. Marshall Ganz, "Resources and Resourcefulness: Strategic Capacity in the Unionization of California Agriculture, 1959–1966," *American Journal of Sociology* 105, no. 4 (2000): 1012.

74. Chattanooga Organized for Action, "People History of Chattanooga," available at https://www.chattaction.org/the-peoples-history-project.html.

75. Alex Press, "The UAW Has Set Its Sights on the Anti-union South," *Jacobin*, March 6, 2024, available at https://jacobin.com/2024/03/uaw-organizing-mercedes-benz-alabama.

76. Jane McAlevey and Abby Lawlor, *Rules to Win by: Power and Participation in Union Negotiations* (Oxford University Press, 2023).

77. Leon, "Southern Autoworkers Organize, Business Class Tries to Wallop Them."

78. *Jacobin Radio*, "UAW's Southern Campaign."

79. Chris Bohner, "Unions Need to Spend Big to Seize the Day," *Jacobin*, May 13, 2024, available at https://jacobin.com/2024/05/union-spending-uaw-workers-united.

80. Because these funds seem to be earmarked for UAW's traditional industrial base, the increase may be even more significant, as the UAW has devoted a significant portion of its organizing budget to academic workers.

81. On the Line, "UAW's Massive Campaign in the South," April 13, 2024, available at https://www.laborradionetwork.org/lrpn-members-1/on-the-line.

82. Luis Feliz Leon, "Twenty-Five Thousand Auto Workers Are Now on Strike at the Big 3," *Labor Notes*, September 29, 2023, available at https://labornotes.org/2023/09/twenty-five-thousand-auto-workers-are-now-strike-big-3.

83. Mike Pare, "Works Council Official Supports UAW in VW Chattanooga Vote," *CTFP*, March 26, 2024, available at https://www.timesfreepress.com/news/2024/mar/26/works-council-official-supports-uaw-in-vw/.

84. Mike Pare, "Global Union Suspends VW Agreement, Cites Chattanooga Plant," *CTFP*, January 22, 2019.

85. Michael Fichter, *Trade Unions in Transformation: Transnationalizing Unions: The Case of the UAW and the IG Metall* (Berlin: Friedrich Ebert Stiftung, 2017).

86. Anonymous informant, interview with author, September 21, 2016.

87. Sigmund Freud, *The Complete Psychological Works of Sigmund Freud*, vol. 23, new ed. (London: Vintage, 2001).

88. Cornelius Castoriadis, *The Imaginary Institution of Society* (Cambridge, MA: MIT Press, 1987).

CHAPTER 8

1. I am détourning James Scott's *Seeing like a State: How Certain Schemes to Improve the Human Condition Have Failed* (New Haven, CT: Yale University Press, 1999). See also Guy Debord and Gil J. Wolman, "A User's Guide to Détournement," *Les Lévres Neus* 8 (May 1956), available at https://www.sionline.researche-editions.cddc.vt.edu/presitu/usersguide.html.

2. Paul Marginson, Keith Sisson, and James Arrowsmith, "Between Decentralization and Europeanization: Sectoral Bargaining in Four Countries and Two Sectors," *European Journal of Industrial Relations* 9, no. 2 (July 1, 2003): 163–87, available at https://doi.org/10.1177/0959680103009002003.

3. Kevin R. Cox and Andrew Mair, "Locality and Community in the Politics of Local Economic Development," *Annals of the Association of American Geographers* 78, no. 2 (1988): 307–25; Nathan Lillie and Miguel Martínez Lucio, "Rollerball and the Spirit of

Capitalism: Competitive Dynamics within the Global Context, the Challenge to Labour Transnationalism, and the Emergence of Ironic Outcomes," *Critical Perspectives on International Business* 8, no. 1 (January 1, 2012): 74–92, available at https://doi.org/10.1108/1 7422041211197576.

4. Rebecca A. Johns, "Bridging the Gap between Class and Space: U.S. Worker Solidarity with Guatemala," *Economic Geography* 74, no. 3 (1998): 252–71, available at https://doi.org/10.2307/144376.

5. Ingemar Lindberg, "Varieties of Solidarity: An Analysis of Cases of Worker Action across Borders," in *Global Restructuring, Labour and the Challenges for Transnational Solidarity*, ed. Andreas Bieler and Ingemar Lindberg (New York: Routledge, 2010), 220.

6. Armel Brice Adanhounme and Christian Levesque, "Southern Perspectives on Transnational Unionism: From the Global to the Local or from the Local to the Global?" *Relations Industrielles/Industrial Relations* 68, no. 2 (2013): 239–61.

7. Christa J. Olson, "Performing Embodiable Topoi: Strategic Indigeneity and the Incorporation of Ecuadorian National Identity," *Quarterly Journal of Speech* 96 (August 1, 2010): 300–323, available at https://doi.org/10.1080/00335630.2010.499108.

8. David Harvey, *Spaces of Capital* (New York: Routledge, 2001), 20.

9. Richard Hyman, "Trade Unions and the Politics of the European Social Model," *Economic and Industrial Democracy* 26, no. 1 (February 1, 2005): 9–40, available at https://doi.org/10.1177/0143831X05049401.

10. Andrew Herod, "The Production of Scale in United States Labour Relations," *Area* 2, no. 1 (1991), 82–88; Andrew Herod, "From a Geography of Labor to a Labor Geography: Labor's Spatial Fix and the Geography of Capitalism," *Antipode* 29, no. 1 (1997): 1–31; Andrew Herod, "Workers, Space, and Labor Geography," *International Labor and Working Class History* 64, no. 64 (2003): 112–38, available at https://doi.org/10.1017/S014754 790300022X; Andrew Herod, "Labor Internationalism and the Contradictions of Globalization: Or, Why the Local Is Sometimes Still Important in a Global Economy," *Antipode* 33, no. 3 (July 1, 2001): 407–26, available at https://doi.org/10.1111/1467-8330.001 91.

11. Katy Fox-Hodess, "(Re-)Locating the Local and National in the Global: Multiscalar Political Alignment in Transnational European Dockworker Union Campaigns," *British Journal of Industrial Relations* 55, no. 3 (2017): 626–47, available at https://doi.org /10.1111/bjir.12222.

12. Andrew Herod, *Labor Geographies: Workers and the Landscapes of Capitalism* (New York: Guilford, 2001), 422.

13. Mario Tronti, "The Strategy of Refusal," Libcom.org, July 23, 2005, available at https://libcom.org/article/strategy-refusal-mario-tronti.

14. Carrie Hertz, "Finding the Local in the Global in the 21st Century," *Practicing Anthropology* 37, no. 3 (July 1, 2015): 56–56, available at https://doi.org/10.17730/0888 -4552-37.3.56.

15. Michael Burawoy, "From Polanyi to Pollyanna: The False Optimism of Global Labor Studies," *Global Labour Journal* 1, no. 2 (2010): 301–13, available at https://doi.org /10.15173/glj.v1i2.1079.

16. Gay Seidman, *Beyond the Boycott: Labor Rights, Human Rights, and Transnational Activism*, American Sociological Association's Rose Series in Sociology (New York: Russell Sage Foundation, 2007).

17. The largely discredited contact hypothesis made a similar point. Gordon W. Allport, *The Nature of Prejudice* (Reading, MA: Doubleday Anchor, 1958).

18. Chris Baldry et al., "Fighting Multinational Power: Possibilities, Contradictions," *Capital and Class* 7, no. 2 (June 1, 1983): 157–67, available at https://doi.org/10.1177 /030981688302000107; Dale Hathaway, *Allies across the Border* (Cambridge, MA: South End, 2000); Nigel Haworth and Harvie Ramsay, "Matching the Multinationals: Obstacles to International Trade Unionism," *International Journal of Sociology and Social Policy* 6, no. 2 (January 1, 1986): 55–82, available at https://doi.org/10.1108/eb013008.

19. Tamara Kay, *NAFTA and the Politics of Labor Transnationalism* (Cambridge, UK: Cambridge University Press, 2011).

20. David Featherstone, "Spatialities of Transnational Resistance to Globalization: The Maps of Grievance of the Inter-continental Caravan," *Transactions of the Institute of British Geographers* 28, no. 4 (2003): 404–21, available at https://doi.org/10.1111/j.00 20-2754.2003.00101.x.

21. David Harvey, *The Limits to Capital* (London: Verso, 2018).

22. See Jamie K. McCallum, *Global Unions, Local Power: The New Spirit of Transnational Labor Organizing* (Ithaca, NY: Cornell University Press, 2013). Still others like Neil Smith have written of "jumping scale," but in the last instance, these models often prioritize large-scale outcomes over their small-scale origins—all the while refusing to interrogate scalarity as such. See Neil Smith, "Geography, Difference and the Politics of Scale," in *Postmodernism and the Social Sciences*, ed. Joe Doherty, Elspeth Graham, and Mo Malek (London: Palgrave Macmillan UK, 1992).

23. Lionel Obadia, "Technocultures and Glocalization," *Glocalism: Journal of Culture, Politics and Innovation*, no.1 (2022), available at https://doi.org/10.12893/gjcpi.2022.1.9.

24. James L. Peacock, *Grounded Globalism: How the U.S. South Embraces the World* (Athens: University of Georgia Press, 2010).

25. John Allen, "Three Spaces of Power: Territory, Networks, plus a Topological Twist in the Tale of Domination and Authority," *Journal of Power* 2, no. 2 (August 1, 2009): 197–212, available at https://doi.org/10.1080/17540290903064267; John Allen, "Topological Twists: Power's Shifting Geographies," *Dialogues in Human Geography* 1, no. 3 (November 1, 2011): 283–98, available at https://doi.org/10.1177/2043820611421546; John Allen, *Topologies of Power: Beyond Territory and Networks* (Abingdon: Routledge, 2016), 201.

26. John Allen, *Power and Space: Essays on a Shifting Relationship* (Abingdon: Routledge, 2024), 156.

27. John Allen, *Topologies of Power*, 51.

28. Phil Williams, "Exclusive: Leaked Emails Raise Questions about Governor's $300M Offer," News Channel 5, April 24, 2014, available at https://www.newschannel5 .com/news/newschannel-5-investigates/tennessees-secret-deals/exclusive-leaked-emails -raise-questions-about-governors-300m-offer.

29. Eric Blanc, "Worker-to-Worker Organizing Goes Viral," *New Labor Forum* 33, no. 1 (2024): 77–83, available at https://doi.org/10.1177/10957960231220914.

30. Dan Clawson, *The Next Upsurge: Labor and the New Social Movements* (Ithaca, NY: Cornell University Press, 2018).

31. Jane F. McAlevey, *No Shortcuts: Organizing for Power in the New Gilded Age*, illustrated ed. (New York: Oxford University Press, 2016).

32. Mark Engler and Paul Engler, *This Is an Uprising: How Nonviolent Revolt Is Shaping the Twenty-First Century* (New York: Bold Type Books, 2016), 185.

33. Daniel Boguslaw, "Big Three Automakers' Reputations Plummet as UAW Strike Rages," The Intercept, September 26, 2023, available at https://theintercept.com/2023/09 /26/uaw-strike-big-three-reputation/.

34. Brooks, "How Amazon and Starbucks Workers Are Upending the Organizing Rules."

35. Brooks, "How Amazon and Starbucks Workers Are Upending the Organizing Rules."

36. Brooks, "How Amazon and Starbucks Workers Are Upending the Organizing Rules."

37. Robin D. G. Kelley, "'We Are Not What We Seem': Rethinking Black Working-Class Opposition in the Jim Crow South," *Journal of American History* 80, no. 1 (1993): 75–112, available at https://doi.org/10.2307/2079698; James C. Scott, *Domination and the Arts of Resistance: Hidden Transcripts* (New Haven, CT: Yale University Press, 1990).

CONCLUSION

1. Lindsay Chappell, "VW's New Tennessee Factory Will Have Key Suppliers on Site," *Automotive News: Detroit* 84, no. 6399 (February 15, 2010): 16B.

2. Lindsay Chappell, "VW U.S. Plant Wants 80% Local Content," *Automotive News: Detroit* 83, no. 6364 (June 15, 2009): 21.

3. Mike Pare, "Volkswagen Chattanooga Supplier Park Marks 10 Years, Readies for Electric SUV," *CTFP*, October 7, 2020, available at https://www.timesfreepress.com/news/2020/oct/07/volkswagen-chattanooga-supplier-park-marks-10-year/.

4. Nathan Bomey, "Volkswagen Supplier Adding 500 Jobs in Chattanooga," *USA Today*, June 23, 2015, available at https://www.usatoday.com/story/money/2015/06/23/volkswagen-chattanooga-gestamp-supplier-expansion/29165285/; Mike Pare, "Volkswagen Supplier to Build $42 Million Plant in Chattanooga, Employ 240," *CTFP*, February 4, 2021, 240, available at https://www.timesfreepress.com/news/2021/feb/04/volkswagen-supplier-build-42-million-plant-chattan/.

5. Mike Pare, "Chattanooga Gives Property Tax Break to Battery Materials Company," *CTFP*, July 6, 2021, available at https://www.timesfreepress.com/news/2021/jul/06/chattanooga-gives-property-tax-break-battery-mater/; Marcus Williams, "VW Group Looks for Partnerships to Build a Resilient North American Supply Chain," Automotive Logistics, December 7, 2023, available at https://www.automotivelogistics.media/supply-chain-management/vw-group-looks-for-partnerships-to-build-a-resilient-north-american-supply-chain/45003.article.

6. Richard Hyman, *Strikes*, 4th ed. (Basingstoke: Palgrave Macmillan, 1989).

7. Mike Pare, "VW Chattanooga, Area Employers Face New Labor Future," *CTFP*, April 20, 2024, available at https://www.timesfreepress.com/news/2024/apr/20/vw-chattanooga-area-employers-face-new-labor/.

8. John Irwin, "UAW Organizes Magna Seating Plant in Tennessee," *Automotive News*, August 25, 2016, available at https://www.autonews.com/article/20160825/OEM01/160829919/uaw-organizes-magna-seating-plant-in-tennessee; John Lippert, "Suppliers Are the Target: UAW Drive to Focus on Outside Parts Workers," WardsAuto, December 21, 2000, available at https://www.wardsauto.com/news-analysis/suppliers-are-target-uaw-drive-focus-outside-parts-workers.

9. Mark Huson and Dhananjay Nanda, "The Impact of Just-in-Time Manufacturing on Firm Performance in the US," *Journal of Operations Management* 12, no. 3–4 (1995): 297–310, available at https://doi.org/10.1016/0272-6963(95)00011-G.

10. Sunil Chopra and ManMohan S. Sodhi, "Managing Risk to Avoid Supply-Chain Breakdown," *MIT Sloan Management Review*, October 15, 2004, available at https://sloanreview.mit.edu/article/managing-risk-to-avoid-supplychain-breakdown/.

11. Daniel Kern et al., "Supply Risk Management: Model Development and Empirical Analysis," *International Journal of Physical Distribution and Logistics Management* 42, no. 1 (January 1, 2012): 60–82, available at https://doi.org/10.1108/09600031211202472.

12. Yossi Sheffi and Jr Rice James, "A Supply Chain View of the Resilient Enterprise," *MIT Sloan Management Review* 47 (September 1, 2005), 41–48.

13. Ila Manuj and John T. Mentzer, "Global Supply Chain Risk Management," *Journal of Business Logistics* 29, no. 1 (2008): 133–55, available at https://doi.org/10.1002/j.2158 -1592.2008.tb00072.x.

14. Mark Brenner, "Union Faces Fresh Questions in West Coast Longshore Standoff," *Labor Notes*, January 22, 2015, available at https://labornotes.org/2015/01/union-faces -fresh-questions-west-coast-longshore-standoff; Kevin O'Riley, "White Paper: Lessons from the West Coast Port Closures of 2014," MD Logistics, June 25, 2015, available at https://www.mdlogistics.com/2015/06/white-paper-the-perfect-storm-lessons-from -the-west-coast-port-closures-of-2014/.

15. Howard Botwinick, *Persistent Inequalities: Wage Disparity under Capitalist Competition* (Boston: Brill, 2018).

16. Jane Slaughter, "Strikes at Walmart Warehouses Expose Threats in Supply Chain," *Labor Notes*, September 24, 2012, available at https://labornotes.org/2012/09/strikes -walmart-warehouses-expose-threats-supply-chain.

17. Ian Thibodeau and Nora Naughton, "Ford Halts F-150 Production over Parts Shortage," *Detroit News*, May 9, 2018, available at https://www.detroitnews.com/story /business/autos/ford/2018/05/09/ford-trucks-production-halted/34718903/.

18. Lisa Xu, "Big 3 Focus on Auto Parts Centers in Strike Prep," *Labor Notes*, September 11, 2023, available at https://labornotes.org/2023/09/big-3-focus-auto-parts-centers -strike-prep.

19. Luis Feliz Leon, "Auto Workers Direct Momentum toward Organizing Plants across the U.S.," *Labor Notes*, November 30, 2023, available at https://labornotes.org /2023/11/auto-workers-direct-momentum-toward-organizing-plants-across-us.

20. Kim Moody, "Modern Capitalism Has Opened a Major New Front for Strike Actions: Logistics," *In These Times*, January 24, 2018, available at https://inthesetimes.com /article/capitalism-logistics-strike-unions-labor.

21. Marco Habermann, Jennifer Blackhurst, and Ashley Metcalf, "Keep Your Friends Close? Supply Chain Design and Disruption Risk" (SSRN Scholarly Paper, Rochester, NY, June 1, 2015), available at https://papers.ssrn.com/abstract=2757754.

22. Yoshio Takahashi and Chester Dawson, "Japan's Big Three Car Makers Poised to Surge," *Wall Street Journal*, April 22, 2012, available at http://online.wsj.com/article/SB1 0001424052702304331204577353442942113590.html.

23. Manuj and Mentzer, "Global Supply Chain Risk Management."

24. Jennifer Blackhurst, Kaitlin S. Dunn, and Christopher W. Craighead, "An Empirically Derived Framework of Global Supply Resiliency," *Journal of Business Logistics* 32, no. 4 (2011): 374–91, available at https://doi.org/10.1111/j.0000-0000.2011.01032.x.

25. Jennifer Blackhurst et al., "An Empirically Derived Agenda of Critical Research Issues for Managing Supply-Chain Disruptions," *International Journal of Production Research* 43, no. 19 (October 1, 2005): 4067–81, available at https://doi.org/10.1080/0020 7540500151549.

26. Blackhurst, Dunn, and Craighead, "Empirically Derived Framework of Global Supply Resiliency."

27. Habermann, Blackhurst, and Metcalf, "Keep Your Friends Close?"

28. Christiaan Hetzner, "U.S. Loses out to Canada for Volkswagen's Billion-Dollar Battery Factory Site," Yahoo Finance, March 13, 2023, available at https://finance.yahoo .com/news/u-loses-canada-volkswagen-billion-155624227.html.

29. Jack Ewing, "Volkswagen Leans on Electric Vehicles and Nostalgia to Grow in U.S.," *New York Times*, February 20, 2024, sec. Business, available at https://www.ny times.com/2024/02/20/business/vw-electric-vehicles-us.html.

30. Jack Ewing, "Volkswagen Will Invest Up to $5 Billion in Rivian," *New York Times*, June 25, 2024, sec. Business, available at https://www.nytimes.com/2024/06/25/busi ness/volkswagen-rivian-vw-investment.html.

31. Gabriel Friedman, "VW Hones In on Manufacturing EV 'Battery of the Future' in Canada," *Financial Post*, updated April 24, 2024, available at https://financialpost .com/commodities/vw-plans-to-make-solid-state-batteries-ontario.

32. Liam Casey, "Here's How Canada Locked Down Volkswagen's First Overseas EV Battery Plant," Global News, updated August 10, 2023, available at https://globalnews.ca /news/9697347/volkswagen-canada-ev-deal/.

33. William Boston, "VW Is on a Hunt for Resources to Remove China from Its EV Batteries," *Wall Street Journal*, June 4, 2023, sec. Business, available at https://www.wsj.com /articles/vw-is-on-a-hunt-for-resources-to-remove-china-from-its-ev-batteries-663ee99.

34. "Manufacturers Pack Meeting for Crack at Supplying St. Thomas VW Plant," *Windsor (Ontario) Star*, April 11, 2024, available at https://windsorstar.com/news/local -news/manufacturers-pack-meeting-for-crack-at-supplying-st-thomas-vw-plant.

35. Luke Martin, "Volkswagen Reorganises Semiconductor Procurement," *Just Auto* (blog), August 25, 2023, available at https://www.just-auto.com/news/volkswagen-reor ganises-semiconductor-procurement-strategy/.

36. Reuters, "Some Volkswagen Cars Delayed in U.S. Ports over Chinese Part," February 14, 2024, sec. Autos and Transportation, available at https://www.reuters.com/busi ness/autos-transportation/some-volkswagen-cars-delayed-us-ports-over-chinese-part -2024-02-14/.

37. Dan DiMaggio, "Big 3 Buckled as Stand-Up Strike Spread," *Labor Notes*, October 31, 2023, available at https://labornotes.org/2023/10/big-3-buckled-stand-strike-spread.

38. Keith Brown, "As Big 3 Auto Contracts Expire: Hurried Line Speeds and Horrible Hours," *Labor Notes*, July 25, 2023, available at https://labornotes.org/2023/07/big-3-au to-contracts-expire-hurried-line-speeds-and-horrible-hours.

39. Megan Kelly, "Battery Supply Chain Could Bottleneck in 2025, Says Analyst," Automotive Logistics, June 01, 2023, available at https://www.automotivelogistics .media/battery-supply-chain/battery-supply-chain-could-bottleneck-in-2025-says-ana lyst/44283.article.

40. Keith Naughton, "Auto Union Blasts $9.2 Billion US Loan to Ford for Creating 'Low-Road Jobs,'" Bloomberg, June 23, 2023, available at https://www.bloomberg.com/news /articles/2023-06-23/union-blasts-9-2-billion-ford-loan-for-creating-low-road-jobs.

41. Jeanne Whalen, "Electric Vehicle Skepticism from Auto Workers Puts Biden in a Tough Spot," *Washington Post*, July 10, 2023, available at https://www.washingtonpost .com/business/2023/07/10/biden-ev-uaw-labor-union/.

42. Jeffrey S. Rothstein, "Lean Times: The UAW Contract and the Crisis of Industrial Unionism in the Auto Industry," *New Labor Forum* 17, no. 2 (2008): 60–69.

43. Adrien Thomas and Nadja Doerflinger, "Trade Union Strategies on Climate Change Mitigation: Between Opposition, Hedging and Support," *European Journal of Industrial Relations* 26 (October 8, 2020), available at https://doi.org/10.1177/0959680120951700.

44. Mathieu Dupuis et al., "A Just Transition for Auto Workers? Negotiating the EV Transition in Germany and North America," *ILR Review*, May 8, 2024, available at https://doi.org/10.1177/00197939241250001.

45. Dana Frank, *Buy American: The Untold Story of Economic Nationalism* (Boston: Beacon, 2000).

46. Luis Feliz Leon, "Auto Workers Spare Big 3, Win Landmark Just Transition at General Motors," *Labor Notes*, October 6, 2023, available at https://labornotes.org/2023/10/auto-workers-spare-big-3-win-landmark-just-transition-general-motors.

47. Kate Aronoff, "Electric Vehicles Have Become a Weapon in the War on Autoworkers," *New Republic*, September 13, 2023, available at https://newrepublic.com/article/175507/uaw-electric-vehicles-strike-workers.

48. United States labor law draws a strict distinction between mandatory and permissive bargaining topics, with implications for economic strikes. See John Thomas Delaney and Donna Sockell, "The Mandatory-Permissive Distinction and Collective Bargaining Outcomes," *ILR Review* 42, no. 4 (1989): 566–83, https://doi.org/10.1177/001979398904200407.

49. Christian Lévesque and Gregor Murray, "Understanding Union Power: Resources and Capabilities for Renewing Union Capacity," *Transfer: European Review of Labour and Research* 16, no. 3 (August 1, 2010): 333–50, available at https://doi.org/10.1177/1024258910373867.

50. Oliver Ibert et al., "Geographies of Dissociation: Value Creation, 'Dark' Places, and 'Missing' Links," *Dialogues in Human Geography* 9, no. 1 (March 1, 2019): 43–63, available at https://doi.org/10.1177/2043820619831114.

51. Richard Hyman, "Imagined Solidarities: Can Trade Unions Resist Globalization?" in *Globalization and Labour Relations*, ed. Peter Leisink (Cheltenham, UK: Edward Elgar, 1999), 94–115.

52. Rebecca A. Johns, "Bridging the Gap between Class and Space: U.S. Worker Solidarity with Guatemala," *Economic Geography* 74, no. 3 (1998): 252–71, available at https://doi.org/10.2307/144376.

53. Johns, "Bridging the Gap between Class and Space."

54. Rochelle DuFord, *Solidarity in Conflict: A Democratic Theory* (Stanford, CA: Stanford University Press, 2022).

55. Emile Durkheim, *The Division of Labor in Society* (New York: Simon and Schuster, 1997), 117.

56. Antonio Gramsci, *Selections from Political Writings: 1921–1926*, reprint ed. (Minneapolis: University of Minnesota Press, 1990), 544.

57. Doug McAdam, Sidney Tarrow, and Charles Tilly, *Dynamics of Contention* (Cambridge: Cambridge University Press, 2001).

58. Sebastian Garbe, *Weaving Solidarity: Decolonial Perspectives on Transnational Advocacy of and with the Mapuche* (transcript Verlag, 2022), 24.

59. Werner Olle and Wolfgang Schoeller, "World Market Competition and Restrictions upon International Trade-Union Policies," *Capital and Class* 1, no. 2 (June 1, 1977): 56–75, available at https://doi.org/10.1177/030981687700200103.

60. Charles V. Hamilton and Kwame Ture, *Black Power: Politics of Liberation in America* (New York: Knopf Doubleday, 2011).

61. Janet M. Conway, Michal Osterweil, and Elise Thorburn, "Theorizing Power, Difference and the Politics of Social Change: Problems and Possibilities in Assemblage Thinking," *Studies in Social Justice* 12, no. 1 (2018): 6, available at https://doi.org/10.26522/ssj.v12i1.1745.

62. Jodi Dean, *Solidarity of Strangers: Feminism after Identity Politics* (Berkeley: University of California Press, 1996).

63. David Featherstone, *Solidarity: Hidden Histories and Geographies of Internationalism* (London: Zed Books, 2012), 63.

64. Sebastian Garbe, *Weaving Solidarity*, 63.

65. Featherstone, *Solidarity*, 38.

66. Featherstone, *Solidarity*, 43.

67. Rick Fantasia, *Cultures of Solidarity: Consciousness, Action, and Contemporary American Workers* (Berkeley: University of California Press, 1989).

68. Fantasia, *Cultures of Solidarity*, 131.

EPILOGUE

1. Payday Report, "Podcast: Payday Report with 70% UAW Sign Up, 25-Year Mercedes UAW Veteran Talks," May 17, 2024, available at https://paydayreport.com/podcast-payday-report-with-70-uaw-sign-up-25-year-mercedes-uaw-veteran-talks/.

2. Luis Feliz Leon, "In Relay Race to Organize the South, Volkswagen Workers Pass the Baton to Mercedes Workers," *Labor Notes*, April 30, 2024, available at https://labornotes.org/2024/04/relay-race-organize-south-volkswagen-workers-pass-baton-mercedes-workers.

3. Timothy J. Minchin, "A World Class Industry: Honda in Alabama and the Rise of the Foreign-Owned Auto Sector," *Journal of Contemporary History* 56, no. 3 (July 1, 2021): 789–816, available at https://doi.org/10.1177/0022009420925883.

4. Dawn Kent Azok, "German Labor Union Playing Key Role in UAW's Latest Campaign at Alabama's Mercedes Plant," AL.com, August 24, 2013, available at https://www.al.com/business/2013/08/post_63.html.

INDEX

Abe Walker is Assistant Professor of Sociology at Fayetteville State University.